JUL 1 9 2013

530.092 Mansf

Mansfield, P.
The long road to Stockholm.

PRICE: $44.95 (3559/he   )

THE LONG ROAD TO STOCKHOLM

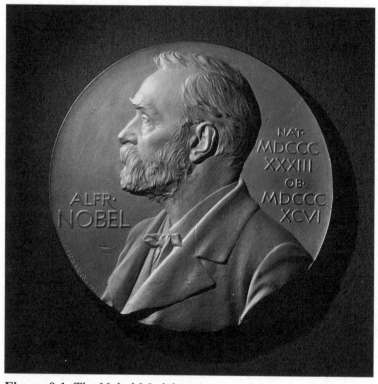

Figure 0.1 The Nobel Medal.
© ® The Nobel Foundation. Photo: © Ted Spiegel/CORBIS.

THE LONG ROAD TO STOCKHOLM

The Story of Magnetic
Resonance Imaging (MRI):
An Autobiography

by

Sir Peter Mansfield

OXFORD
UNIVERSITY PRESS

Great Clarendon Street, Oxford ox2 6DP,
United Kingdom

Oxford University Press is a department of the University of Oxford.
It furthers the University's objective of excellence in research, scholarship,
and education by publishing worldwide. Oxford is a registered trade mark of
Oxford University Press in the UK and in certain other countries

© Sir Peter Mansfield 2013

The moral rights of the author have been asserted

First Edition published in 2013

Impression: 1

All rights reserved. No part of this publication may be reproduced, stored in
a retrieval system, or transmitted, in any form or by any means, without the
prior permission in writing of Oxford University Press, or as expressly permitted
by law, by licence or under terms agreed with the appropriate reprographics
rights organization. Enquiries concerning reproduction outside the scope of the
above should be sent to the Rights Department, Oxford University Press, at the
address above

You must not circulate this work in any other form
and you must impose this same condition on any acquirer

British Library Cataloguing in Publication Data
Data available

ISBN 978–0–19–966454–2

Printed and bound by
CPI Group (UK) Ltd, Croydon, CR0 4YY

Oxford University Press makes no representation, express or implied, that the
drug dosages in this book are correct. Readers must therefore always check
the product information and clinical procedures with the most up-to-date
published product information and data sheets provided by the manufacturers
and the most recent codes of conduct and safety regulations. The authors and
the publishers do not accept responsibility or legal liability for any errors in the
text or for the misuse or misapplication of material in this work. Except where
otherwise stated, drug dosages and recommendations are for the non-pregnant
adult who is not breast-feeding

Acknowledgements

I am most grateful to the following friends, colleagues, and referees for critical and helpful comments and suggestions on this book: Roger Ordidge, Ian Pykett, Penny Gowland, Barry Chapman, and Brett Haywood.

I wish to thank my secretary, Pamela Davies, for the unstinting help and patience she has shown to me in the course of the preparation and retyping of the many versions of the manuscript, and Elizabeth Croal for her help and assistance throughout all phases in the manuscript preparation.

I also wish to record my special thanks and appreciation to my wife, Jean, and my two daughters, Sarah and Gillian, for their forbearance and understanding during the course of my long and active career.

Contents

Abbreviations ix

1. The War Years 1
2. Rocket Science 27
3. Salad Days at University 47
4. Marriage and the American Dream 63
5. The Wanderers Return 77
6. Alt-Heidelberg 89
7. Krakow and the Lauterbur Epiphany 109
8. Hounsfield and EMI 115
9. The Golden Years 129
10. Brief Encounter with Technicare 145
11. Patent Affairs at Nottingham 155
12. A New Vice Chancellor 165
13. Nobel Prize Speculation 181
14. Antagonisms to MRI 193
15. Beyond the Nobel 201
16. The Epilogue 211

Appendix 227
Index 231

Abbreviations

ANMR	Advance Nuclear Magnetic Resonance
BA	British Association
BOD	Base Ordnance Depot
BP	British Petroleum
BRSG	British Radio Spectroscopy Group
BTG	British Technology Group
CAA	Civil Aviation Authority
COD	Command Ordnance Depot
CT	computerized tomography
ENC	Experimental NMR Conference
EPI	echo planar imaging
ESR	electric spin resonance
EVI	echo volumar imaging
fMRI	functional magnetic resonance imaging
GE	General Electric
GfR	Die Gesellschaft für Raumschiffahrt
HNC	Higher National Certificate
ISMAR	International Society for Magnetic Resonance
JCAT	*Journal of Computer Assisted Tomography*
k-space	reciprocal lattice space
MIT	Massachusetts Institute of Technology
MR	magnetic resonance
MRC	Medical Research Council
MR Centre	Magnetic Resonance Centre
MRI	magnetic resonance imaging
NIH	National Institutes of Health
NMR	nuclear magnetic resonance
NPL	National Physical Laboratory
NRDC	National Research and Development Corporation

PPL	private pilot's licence
PPLH	private pilot's licence, helicopter
r-space	real space
RAOC	Royal Army Ordnance Corps
RASC	Royal Army Service Corps
RCR	Royal College of Radiology
RF	radio frequency
ROF	Royal Ordnance Factory
RPD	Rocket Propulsion Department
SMRM	Society of Magnetic Resonance in Medicine
SRC	Science Research Council
WOSB	War Office Selection Board

1

The War Years

In the Beginning

I was born on 9 October 1933 in Lambeth, London, the youngest of three sons, and grew up in Camberwell, ten minutes' walk from Camberwell Green, the epicentre of the borough. My father, Sidney George Mansfield, was the eldest son in a family of five sons and four daughters, and his parents, my grandparents, were Harry and Mary Mansfield. I do not recall much about Mary, who died in August 1935 aged 58, but I have some vague memories of Harry, a horse trader and dealer in horse harness, who was killed in an air raid early on in the war.

My father worked initially as a labourer in the coke house for the South Metropolitan Gas Company, but later, after he had gained some experience, as a gas fitter. My mother, Rose Lillian, was the youngest of three daughters and as a child lived in Ponsonby Place, just on the north side of the Thames by Vauxhall Bridge. She and her sisters would often visit the street market in The Cut by Waterloo Bridge.

My grandparents on my mother's side were Jack and Margaret Turner. Margaret (see Figure 1) died in a road accident in the early 1930s, and I have no recollection of her at all, but my grandfather Jack (see Figure 2) was a tall upright man who had served in the First World War as a sergeant major, and even out of uniform he had a military bearing and the air of a strict disciplinarian.

My mother had worked as a waitress in Lyons Corner House in the West End, and had also worked as a waitress

Figure 1 My maternal grandmother, Margaret Turner.

during the Crystal Palace Exhibition in 1920. However, she was not working at the time of my birth but stayed at home and looked after me and my two brothers, Conrad William, the eldest, who was nine years older than me, and Sidney Albert, who was six years my senior. They were referred to within the family as Connie and Sidey respectively.

I have only a vague recollection of events when I was very young. I remember a visit to Southend with my parents—I would have been 3 or 4 years old at the time (see Figure 3). My father won a prize on a coconut shy and chose a papier-maché doll of Popeye, filled with toffees, which was presented to me. I also recall the coronation of King George VI

Figure 2 My maternal grandfather, Jack Turner.

in 1937, when there was a street party with flags and bunting criss-crossing the road.

It was around about then that my mother sent Sid and me off for a short country holiday paid for by what was called the Children's Country Holidays Fund, now renamed CCHF All About Kids, a charity that was founded to take disadvantaged London children for fresh-air breaks to the country or seaside. The holiday was a week in Kent, where we all stayed in a large house (see Figures 4 and 5). The event was organized by Mr Cairnes, headmaster of our local Cork Street School, and we were accompanied by one or two teachers. My parents had given me a little spending money, and I had

Figure 3 With Mum and Dad in Southend, 1937–8.

decided to use an old fish paste jar as a moneybox. Unfortunately it hadn't been washed out so that after a few days the odour of rotting fish followed me around, something to which I was completely oblivious, but others weren't, and soon pointed it out to me, somewhat bluntly!

On 3 September 1939, when war broke out, I was only 5 years old, so my memory of much of this period is now somewhat hazy. However, I do recall very clearly the first time the air raid siren sounded on the day war was declared. I was playing in the street nearby and ran home asking what the strange wailing sound was.

In the months that followed plans were instigated to evacuate all children from the London area. Although we were unaware of the dangers at the time, the British Expeditionary Forces were in full retreat and being evacuated from Dunkirk as an emergency towards the end of May 1940. Initially young children were sent to areas close to London. My brother

Figure 4 Country holiday group in Kent, 1938. I am in the centre, with Sid behind me on the back row.

Sid and I, now 12½ and 6½, were evacuated to Sevenoaks, Kent, initially to live with a family that already had children. However, the extra burden imposed on the family by my brother and me was clearly too much for them, so we were transferred to another family with no children. The lady of the house was youngish and terribly house proud and as a result our lives were made extremely difficult—children and being house proud just do not go together.

Two events stick in my mind from this time. The first was when my brother and I were playing with some friends in an old derelict brickyard. Several of the clay pits had sprung

Figure 5 Sidney with a fishing rod in Kent, 1938.

leaks and were filled with water. There was an old rail track by one of these ponds, and on this track was a small handcart. I was pushing the cart and slipped on the track. By then the cart, which was filled with other children, was moving down a slight gradient. I continued to grip the cart and was dragged several yards. My new pullover was ripped, but luckily I was unhurt myself. The second event was a bout of severe constipation, no doubt caused by the trauma of living in this household, and in an effort to relieve my condition I was overdosed with liquid paraffin. This passed straight through me with unfortunate consequences to both my underpants and my trousers, which was, of course, horribly embarrassing to me. In the end my mother came down to visit and when she learnt of our plight she took us away immediately and returned us to London.

In fact there was very little war activity happening in London so far as I was concerned. But of course this was in the early days of 1940 and things were soon to change dramatically. My eldest brother Conrad, who was then 15 years old, had started work and later he volunteered to serve in the Fleet Air Arm, but Sidney and I were still at school.

The Battle of Britain started late in the afternoon of Saturday, 7 September 1940. I watched waves of German bomber aircraft darkening the sky as they flew overhead on their way to bomb the docks in East London. I was by now nearly 8, so this was almost as exciting as it was frightening. The air raids continued for the rest of that month, mainly over southern England and in daylight, with the Royal Air Force constantly harassing the enemy. This resulted in heavy German losses, and these air raids effectively ended on 30 September 1940, when Germany, having lost this famous battle, decided to take a different approach.

As far as we were concerned, the war really started in 1941, with heavy air raids, mostly at night. During this time we got used to regularly living underground, taking our bedding and belongings each night to the communal air raid shelters. I well remember standing with my father at the entrance of the shelter, looking skywards and watching the anti-aircraft shells exploding around the German bombers caught in crossed search lights. Mobile anti-aircraft guns appeared on our street, firing at the aircraft, and I heard the whine of shrapnel raining down as we watched. Then in the daylight we youngsters would search the streets for pieces as souvenirs. As the blitz worsened, new arrangements for the evacuation of young children were announced, and this time my brother and I were sent much further away, to Torquay in Devon, and my memories of that time are many and varied.

Devon, Glorious Devon

My mother came to Paddington Station to bid us farewell. All the children were labelled with tickets, like living parcels, and we were herded into the train compartments with much excited chatter and many tears (see Figure 6). As we entered the carriage, my mother gave me a whole packet of cream crackers and a box of Kraft cheese spread to eat on our way, and long hours later we finally arrived in St Marychurch, Babbacombe, where we were assembled at the Town Hall for dispersal (see Figure 7). A Captain Faversham was in charge, and my brother and I were allocated accommodation with Mr and Mrs Rowland at 4 Carlton Road, Babbacombe, Torquay.

Figure 6 Model figures of evacuee children, just arrived at the dispersal centre in St Mary church, Babbacombe. Note the labels and gas masks; the models resemble how Sid and I must have looked.

Figure 7 The Town Hall at St Mary's Church, now up for sale.

It was clear from the beginning that we would be very happy and extremely comfortable staying with Florence and Cecil Rowland, two wonderful people in their mid to late fifties who had no children of their own. On the train down we had shared a compartment with the Skinner children,

John, who was the eldest, his sister Miriam, and the youngest, Arthur. Arthur was about 7, which was my age, and John was roughly the same age as my brother Sid. A week after our arrival in Babbacombe John managed to contact Sid and tell him they were very unhappy in their new accommodation and wanted to change billets or go back to London. Sid mentioned this to Mrs Rowland, who made enquiries to see if it were possible for the Skinners to come and join us at 4 Carlton Road. Thankfully it was, and shortly afterwards they joined us, and we all settled happily into our new life in Babbacombe. This, then, was my first stay in Devon, from 1940 to 1943, when the Skinners were living with us (see Figure 8).

Figure 8 Auntie and Uncle with the evacuees, 1940–1. Mr and Mrs Rowland, centre, with John Skinner, back left, and Sidney Mansfield, back right. Front row left to right: Arthur Skinner, Miriam Skinner, and me, otherwise occupied.

The Rowlands asked us to refer to them as Auntie and Uncle, which we were all happy to do, because we felt so at home with them. Uncle was a carpenter and joiner and worked in a small firm in Torquay, but he also had a well-equipped workshop in the garden and enthusiastically demonstrated the carpentry facilities to us boys. He gave me a small wooden toolbox that he had made so that I could start my own tool collection, I saved the sixpence pocket money that was sent regularly by my mother, but, though I was able to buy the odd tool from time to time, the bulk of my tools were given to me by Uncle. Sid also showed some interest in woodwork and acquired a set of plans to build model furniture for a doll's house. One of the items that he constructed was a small table with matching chairs. All were made of plywood and cut out using a fretwork saw. I do not recall what happened to this work, but I do remember that the table and chairs were beautifully made.

Carlton Road joined Bronshill Road, and it was along Bronshill Road that I went to school, where we were housed in temporary huts. However, they were close to home and most convenient. My brother went to the more senior Westhill Road School, which was nearby. Bronshill Road was, I conjectured imaginatively, so named because of the rows of beech trees lining the road. In the autumn when the trees shed their nuts the pavements were completely covered in places with these strange triangular-shaped beech nuts and bronze-coloured leaves. I was, of course, wrong—but then I could not read and I certainly did not know then the difference between 'brons' and 'bronze'! I did learn to read later at the age of 9.

With the exception of birds, animals were not allowed in the flats I had lived in with my parents in south-east London. However, when we went to Devon, Auntie and Uncle actually had a cat called Peter, a large brown furry animal, and I fell in love with him immediately. At the time Peter must have been around 7 years old. He seemed to adapt immediately to the house being full of young children and loved joining in with our various activities. So, for example, I would bring out my toy soldiers and set them up on the carpet in front of the fire. Peter would wait until everything was in place, with the soldiers facing each other in battle-lines. He would then stroll right over the middle of the battlefield, look around, and then turn round and round several times in the middle of my soldiers, knocking them all down. He would then lie down, flattening the whole battlefield and disrupting all the soldiers, the fortress, and all the other things that had been set up. At that point he would look around with a loud purr, announcing that he was there to join in the fun.

Mrs Marian was Auntie's slightly younger sister who lived on the Watcombe Road, just outside Torquay. About three or four miles beyond her house was the then famous Watcombe Potteries, which we visited many times and from which we bought small items of locally thrown pottery made in the typical red Devonian clay. During the whole time that we were in Torquay I suppose we saw Mrs Marian no more than half a dozen times, but we never met her husband. I believe, therefore, that she was widowed, but I can't be sure whether she was when we first met her. How many children pay close attention to adults who enter their lives infrequently? However, I did overhear a conversation once between Mrs Marian and her sister, and she spoke about someone dying, and I have a suspicion that it could well have been her husband.

The harbour in Torquay was protected by an outer harbour wall, so that large waves and choppy water were effectively smoothed out when the waves reached the inner harbour. We would often go down to the harbour and watch the fishing boats unload their catch, seagulls shrieking above. It was possible in those days to buy sprats on the harbour side and for sixpence you could get a big bag full of these. On one occasion my brother bought some and took them home and we all enjoyed eating them after Auntie had fried them. She dipped them in batter and then cooked them in the frying pan and they were delicious.

Not far from the harbour was the Prince's Theatre, which still exists, although it is now called the Princess Theatre. Just alongside the Prince's Theatre there was considerable work going on to extend the harbour area. Virtually the whole time that we were in Torquay, there were big pile driving machines along the outer harbour side, pounding away at these corrugated metal piles, which were driven deep into the sand to strengthen the extended harbour wall.

Union Street is the main street running through the centre of the Torquay shopping area. The bottom part is called Fleet Street, and this branches off into Abbey Road, which runs more or less parallel to Union Street. Union Street leads up to Castle Circus, which was something of a bus terminus, and from there one could catch buses to various places in and around Torquay. In those days buses ran from the harbour area along Union Street up to Castle Circus and one could take the bus from Castle Circus up Windsor Road and eventually to Babbacombe. On the way up the hill to Windsor Road via St Marychurch Road not too far from Castle Circus there was a building to the left as one went up the hill. This was the meeting place of the Boys Brigade, of which I was a member. During my first period in Devon, which lasted from 1941 to 1944, I went regularly to the Boys Brigade sessions during the week, and I very much enjoyed the whole process from initiation to passing the necessary exams to obtain the belt and the Boys Brigade forage cap, the badges, and also the white linen shoulder bag. Among the many things that we learnt there was the origin of our own flag, the Union Jack, and its make-up, and also an elementary knowledge of first aid, which, on many occasions, was extremely useful.

Team spirit and working together were also emphasized, both valuable lessons for life.

From time to time during 1943 and early 1944 Torquay was attacked by hit-and-run raiders—German planes that came over and mounted a surprise attack. In one foray they came over on the Sunday morning of 30 May 1943 and dropped bombs in and around Babbacombe. Sadly some hit St Mary church in Babbacombe, where a group of children were gathered with their Sunday school teachers. More than twenty children were killed in this air raid. One of the young boys killed was a schoolfriend of mine called Donald Hext. He, together with the other children and teachers, are buried in St Marychurch graveyard (see Figure 9). One of the enemy aircraft clipped the spire of the Roman Catholic Church nearby and crashed into houses in Teignmouth Road, killing the pilot and least one other person.

As the war proceeded, a number of hit-and-run raids were made in and around Torquay, and it was decided by the government that those who wanted to have a Morrison shelter, rather than an Anderson[†] shelter, could have it installed. Auntie and Uncle decided to have the Morrison shelter, which was like a very large table made of steel, with four legs and caging material around the sides. The idea was that, if a bomb were to hit the house, the masonry might fall and cover the shelter, but hopefully if you were inside the shelter you would be perfectly safe. I learnt only recently that families were charged £7 for the shelter if their annual income was over £350. As the war went on the Morrison shelter was never used in anger.

[†] I believe the Anderson shelter was made of corrugated mild steel sections and partially let into the ground. The earth removed was then spread over the exposed corrugated section giving about 1 foot of earth covering. These shelters could accommodate up to four persons.

Figure 9 Memorial engraving above the entrance to St Mary church, commemorating the loss of life following the bombing of the church on 30 May 1943.

Cockington Village is a small quaint village nestled between Torquay and Paignton, lying about two or three miles behind the main railway station at Torquay. In the early 1940s it was an attraction, just as it is today, and many people came to see the old village and in particular the forge. The forge was actually operating back then, and we often visited it to watch the large plough horses being shod by the village smithy. In those days it was possible to have tea and cakes in one or two of the shops that had opened in that area. These days, although the forge is still there, it is not operating as such, and the whole area has now become a show piece for trinkets and various memorabilia of the age. Of course in the early 1940s, when we visited Cockington, we had to walk from Babbacombe down into Torquay and then along the sea front past Torre Abbey Sands to Cockington Lane.

I made the trek with the others to go collecting nuts in Cockington. Cockington Lane is a long road that leads from the sea front area, with Cockington Village at the end. On the way to the village we would pass woods with many hazel and chestnut trees, and in the autumn we would go to collect hazelnuts, cobnuts, and chestnuts. I would take a small sack—I think it was an onion sack—and we would fill this to the brim with nuts. It was a struggle to bring these home to Carlton Road, but there we would roast them on a shovel placed over an open fire.

In the weeks leading up to Christmas all the children would start thinking seriously about making presents and Christmas cards. Of course, because we were living away from our parents, who were still in London, our first thoughts were about making suitable Christmas cards decorated with nice patterns to send home to them. But we were also concerned about Auntie and Uncle and what we could make for them. We would make spills for lighting cigarettes and pipes, and tapers for lighting the fire as well as the stove and oven; other favourite gifts included a tobacco pouch, a tin of tobacco, or some cigarettes.

Just off the esplanade on Torquay sea front there is a large open park area, Torre Abbey Meadow, which is close to the main railway station. Further back on this green area is Torre Abbey, and in front of the Abbey there is an area of gardens, where there was a water chute and a long narrow strip of water where one could sail model yachts and boats. Auntie took me several times to this park on a Sunday afternoon so I could sail my model yacht. In fact this water strip, or at least the concrete that held the water strip, is still there, even today, sixty years on, but it is often dry through lack of a continuous flow of water. Of course, Torre Abbey is the place where French and Spanish soldiers and sailors captured from the Spanish Armada were incarcerated for a period during the attempted invasion of Britain during the reign of Queen Elizabeth I.

During the early part of 1943, we could see a glow on the south-western horizon from Torquay, which was a result of the heavy bombing of Plymouth and Devonport. Devonport was a major Naval Base and was therefore a major target for the Germans during the war.

Return to London in 1944

During the latter part of 1943 and the spring of 1944 there was very little enemy activity over London and it was thought safe enough for me to return to live with my parents. There had been major allied victories in North Africa, with the 8th Army under General Montgomery defeating Irwin Rommel in Tobrouk, and the men of the German army in North Africa had been either killed or captured and made prisoners of war. However, during the early summer of 1944 a new threat appeared over England—the unleashing of the new and hitherto secret German weapon, the V-1.

The V-1 attack started on 13 June 1944. The week before, on 6 June 1944, my father had taken me to a military display that was being held in Trafalgar Square in central London. A variety of interesting military equipment was on show, including vehicles with static displays depicting photographs of the fighting that had been going on, mainly in North Africa. A couple of tanks were there, and children could climb into the vehicles and have a look around to inspect all the intricacies of the controls. There was also an anti-aircraft gun, and I climbed on it, turning the handles and adjusting the firing angle. It was all fascinating for children, literally *Boys' Own Annual* material, and I thoroughly enjoyed the experience.

In the middle of the excitement of looking at this display, a paper boy selling the *Star* evening newspaper on the corner of Trafalgar Square started to call out: 'Invasion, read all about it!'

A small crowd quickly gathered round him. My father made his way over and bought a copy of the *Star*. There were the headlines in bold black print: **Allies Land in Normandy**. There was much excitement when we got home, and my father said to me and my mother that: 'Hitler won't let us get away with this. We were going to have more raids. He'll probably start blitzing London again.'

The following week—I believe it was a Sunday evening—my father, who had been in the front room of our flat and had seen what looked like an aircraft on fire going towards central London, called us excitedly. 'Come quickly, look, look at this.'

My mother and I assumed it was an aircraft that had been hit and was about to crash, but my father was adamant that this was no ordinary aircraft; this was something new. Flames could be seen coming out of the back of this aircraft-like-object, and it made a very strange noise, which was quite new to us. It flew over the flats until it was almost out of sight and then we heard a huge explosion. The next day the newspapers reported a government denial that the plane seen the night before was a new secret weapon. They claimed that it was simply an aircraft that had been shot down. But several days later the government was forced to admit that this was indeed a new secret weapon that was being used against us, a flying bomb, one of the first V-1 devices to come over London. The Londoners quickly renamed it the 'Doodlebug' or the 'Buzz Bomb', because of the strange noise that it made, and in the ensuing weeks we saw and heard hundreds of these weapons fly overhead.

On one occasion I was near the front entrance to the block of flats where we lived. It was a brilliant sunny day, and the roof of the three-storey building was casting a shadow that half covered the road outside. As I stood there, I heard a Doodlebug coming, and, to my horror, the engine cut out as I was listening. It was clearly quite close. Panic stricken,

I watched the wing in shadow pass along in front of me down the whole length of Picton Street. I stood in the doorway watching mesmerized as it passed. It must have been about 20 or 30 feet above the roof, quite low. Fortunately it had enough speed and altitude to continue gliding. I heard the swishing of air as it passed overhead, and I ran in to the back of the staircase and crouched down. There was an enormous bang and when I went out and looked afterwards I could see that it had landed at the end of our street and hit a small cluster of houses on Benhill Road, which it completely demolished.

I learnt later, after the war, that the Doodlebug carried a 2,000 lb warhead and that its engine was based on the German Schmidt-Argus duct, which in turn was based on the much earlier French design, the Lorraine duct, originally conceived during the First World War.

As the summer of 1944 went by the Doodlebug raids intensified, but then they slowly decreased, as the anti-aircraft batteries, which had initially been placed in London, were moved out to the coastal areas and as our pilots, flying Spitfires and Hurricanes, learnt to deal with the Doodlebugs, quite often by tipping the wings. I saw several examples of this where a Spitfire would fly alongside a Doodlebug and then the pilot would put the wing of his aircraft under the Doodlebug wing and gently nudge it, which would upset the gyro balance mechanism inside and cause the thing to dive. This turned out to be the best way of dealing with them over open country, but of course over the city, although it was tried once or twice, it really did not help. It merely meant that the Doodlebug did not reach its original target but fell on ordinary houses, so that was quickly abandoned as a technique.

After seven weeks of almost continuous air raids, the Doodlebug threat virtually ceased, except for the odd one or two that got through, and we all began to settle back to more or less normal life. But later that summer, on 8 September 1944, we began to experience a new and terrifying weapon, the V-2. The V stood for the German word *Vergeltungswaffe* or reprisal weapon. These really were terrifying, because you could not see or hear them coming. You heard them only after they had hit the ground and exploded. Then there was a long whooshing of air and a rumbling, which was the sound wave catching up with the vehicle itself. By then, of course, the V-2 had already knocked down houses and killed people. The poor people who remained in London, including both my parents, had to live through this period not knowing what was coming and when.

It was not until 10 November 1944 that Prime Minister Winston Churchill formally admitted that the Germans were bombarding London with this new weapon, the V-2.

Return to Devon

I had left Devon early in 1943 because of a lull in air raids over London. But in October 1944, when I was 11 years old, there was renewed bombing in London, so I had returned to Torquay. I was now too old for the Junior School and so went to Audley Park School, a senior school a mile or so from Carlton Road. I used to walk to school, which was in a large and modern building. I clearly remember the art teacher at Audley Park showing me some pieces of printers' type, the large letters used for headlines in newspapers. I was quite interested in these, and the experience stuck in my mind. It may well be as a result of this event that later on when I got back to London I developed an intense interest in printing.

During this second stay I joined the Sea Scouts. The Sea Scout hut was down by the harbour in Torquay but not actually at the harbour level. One had to go up steps, so that we were 30 feet or so above the harbour. I often rowed with several other boys and the Scoutmaster in a large rowing boat in the inner harbour, but when we rowed to the outer harbour the water was really quite choppy.

In 1944, Sid, who at the time was 17 years old, volunteered to join the Royal Navy (see Figure 10). He was based in Devonport, Plymouth, which is not too far from Torquay, and he came back in his uniform after a period of initial training to see Auntie and Uncle and me. One of the things that I arranged for him was a cinematographic show, because during my return to London between stays in Torquay I had been given a hand-operated 35-mm projector as a Christmas present. I had reels of old film—not complete films—but bits and pieces of material that I had melded together to make a film show.

Figure 10 Sidney Mansfield in naval uniform, 1944.

I had changed the projection lamp, because the original version had a small six volt bulb of some sort and it really was not terribly powerful. I built a box on the back of the projector and put in a large 60W bulb, which, of course, increased the clarity of the projection. It was this arrangement that I had set up in the Morrison shelter, and when Sid arrived the first thing he had to do was sit in the shelter with me while I projected the film programme. I suspect it must have bored him to tears, but he did his part dutifully. And for the privilege I had the audacity to charge him five shillings.

Now that Sid had joined the Navy, and the Skinners, who had also returned to London when I did, decided to stay there, I was left in Devon on my own. Woodwork was still a hobby, and one that was strongly encouraged by Uncle. Any spare money that I managed to save still went towards more woodworking tools, and, with the help of the tools that had been given to me by Uncle, I had managed to build up a useful set. One of the items that I bought was a small metal rabbit plane used for rebating wood. This allows one to plane right up to the inside corners, for example, in an extant frame such as a window sash, and, with Uncle's guidance, I made several toys, including a small model of a truck. It was a 1930s-style truck with an open cab at the front. I built this and painted it in suitable colours, but of course I did not have any machinery, so when it came to the wheels of the truck, Uncle offered to get these turned for me at his firm. Later on I decided that I would try to sell these toys and one Saturday I went into Torquay looking for a suitable toyshop. The first place I tried was close to the undercover market in Market Street and luckily the owner was amenable to acquiring the toys that I had made. I made several others, and they were placed in the window of the shop in Torquay and eventually sold for a few shillings each.

In the run up to the D-Day landings in 1944 there was tremendous military activity in and around Torquay. The heavy

The War Years

tread of men marching could be heard regularly, as well as the rattle of machineguns, especially at night. One day I had been taken by Auntie to Paignton for the afternoon. I remember that while in Paignton I had been mesmerized by some polished ornamental rocks that had been decorated with little models of seagulls. These could be bought for a few shillings, and on this occasion were on display in the shops in Paignton, although they were also available in shops in Babbacombe.

We travelled back to Torquay on a bus, and as we came along the sea front from Paignton into Torquay we reached Corbyn Head, which was at that time occupied with a military installation that included a huge ex-Naval gun pointing out to the sea in order to protect the entrance to the Torbay area.

From my vantage point upstairs at the front of the bus I noticed that there was considerable activity ahead of us. Looking down, I could see streams of tracer bullets crossing the road ahead of us from one side to the other, and it was clear that there were heavy military training manœuvres going on, even as the traffic passed through. Fortunately, we were not hit! Presumably the military held up for a moment or two, while we passed, before continuing. They appeared to be using live ammunition in their activities.

The raids continued from time to time, and in early 1944—it must have been during a school holiday period, because I was playing outside the house in Carlton Road—an aircraft flew over at virtually roof-top height. As I looked up, I saw the yellow painted underbelly of a large twin-engine German plane, a Fokke-Wulf, I believe, with large swastikas and iron cross markings. The plane had a tail gunner, and as it flew over the gunner was spraying bullets everywhere. I ducked down behind a dry-stone wall dividing the back gardens of some adjacent properties and raised my head to watch the plane fly by. It went over my head and continued straight on into Torquay, following pretty well the route of Windsor Road.

This experience really scared the life out of me. A little later that summer, when my parents came to visit me, there was a loud bang as we walked down Market Street. It came from the backfiring of a van or truck, but I immediately ran for cover and crouched down close to the pavement, terrified. My parents were amazed to see me do this. They had thought that Torquay was a safe place for children, and I do not believe they had any real idea there were such dangers. Later on during their holiday visit we went to Dartmouth by train and looked around the quaint and beautiful town. Vividly imprinted on my memory is the recollection of standing on some high ground by the railway station at Kingswear, where we could look down on the river Dart. On the other side of the river we could see the Naval College, and along the river Dart, moored each side of the river, were dozens of invasion barges, as far as the eye could see. These khaki-painted boats were being assembled, as we learnt later, for the Normandy landings.

On another occasion, as I was walking with my parents along Cockington Lane, we heard the sounds of planes and looked up to witness an aerial dogfight in progress between German fighter planes and our own fighters. This was going on right above our heads, with cannons and machine guns being fired. It was really a spectacle, and one that I had not witnessed before at such close quarters. As we stood there looking up in the sky a young air force chap who had been walking along the lane stopped to watch. It turned out that he was a Polish airman, and he stood there with us and explained what the aircraft were. He was very excited watching the planes swooping in and out and burst into Polish—presumably he was urging our fighter pilots to shoot down the enemy.

Another place I visited with my parents was Kent's Cavern is in Wellswood, which is just off the road leading from the harbour area in Torquay up to Babbacombe. It is a natural underground cave, with beautiful stalagmites and stalactites, and in one particular area the cavern opens up into a huge chamber in which all the surrounding rocks glisten when illuminated with light. I was completely fascinated with the place and went on at least two occasions. In one part of the cavern there is evidence of Neolithic Stone Age occupation, and various bones had been uncovered there and some elementary tools found.

Teignmouth lies roughly 10 miles along the coast from Babbacombe. In the 1940s it was a small fishing village, and one day the school organized a day trip for some of the pupils. We gathered together in Babbacombe and walked along the coastal pathways all the way to Teignmouth. I forget now actually how long it took, but by the end of the journey I was absolutely worn out and I felt as if I could not walk another step. We *all* felt very tired. Ten miles, of course, was for young children like walking to the moon, but when we finally got to Teighmouth the teacher organized a bus trip back thus avoiding total exhaustion!

In early 1945 Auntie was taken ill and was confined to her bed for a period of time. It was at this time that I saw her sister, Mrs Marian, once more. She came to look after Auntie and told me that it would not be possible for me to continue to stay in Torquay. My parents were informed and came down to collect me for my return to London. I was sad to leave Auntie and Uncle, and Devon itself. I returned to London when the city was in the middle of a thick, dense fog, or smog as it was called, a mixture of greeny yellow smoke and fog that made it very difficult for some people to breathe. It was so bad in places that it was difficult for the bus drivers to make their way and the conductor on our bus had to walk ahead of the bus making sure there was no obstruction.

After the war, in 1949, the Rowlands moved from Carlton Road, Babbacombe, to St Marychurch. Uncle, Cecil Henry

Coysh Rowland, died in May 1951, aged 69. Auntie, Florence Susan Rowland, later moved to a retirement home in Paignton, where she died in October 1954, aged 70. Both Auntie and Uncle were laid to rest in Torquay cemetery. They had given me a wonderful home and a wealth of experiences and even today I remember them with great affection.

2

Rocket Science

Return to London

By now I was 11 years old and had returned to the Junior section of Cork Street School. One day in the New Year I was approached by my schoolteacher, a Mr Volume, about taking the 11+ examination. He said that it was going to be held fairly soon and I was given something like a week to prepare for this. I had never heard of the 11+ examination before, so this was all quite new to me. I took the exam but failed to pass with sufficient marks to go to the local Grammar School. My pass level was, however, sufficient for me to go to an intermediate school called Peckham Central School.

Peckham Central School was just off the Peckham Road before the intersection of Peckham Road and Sumner Road. In fact there were two schools on the site, Sumner Road School, an elementary school, and Peckham Central School. The two schools were separated by playing areas and a playground.

Peckham Central was a mixed school, and my form master was Mr Tiller. The school itself had a junior section and a senior school with pupils up to the age of 18 years. The last two years of school were spent in the lower and upper sixth forms. As well as a range of ordinary subjects, such as English, Mathematics, and so on, we started to learn French. The French mistress's name was Miss Lethbridge. I continued at this school for about one year, but towards the end of that year we were informed that, because of the 1944 Education Act, Central Schools would cease to exist. There was in fact a huge reorganization of secondary schools in the London area, and Central Schools were merged with ordinary Elementary Schools into a new type of school, which was called a Secondary Modern School. The Secondary Modern would further evolve into what are now called Comprehensive Schools, but this happened long after I had left school in 1948. In our case, the new Secondary Modern School was single sex. After one year at Peckham Central School, the boys were split off and merged with boys from other schools to form the Secondary Modern School at Choumert Road in Peckham. Choumert Road was a mile or two further away from home, so I had quite a walk to and from school. I can just about remember the names of one or two of the teachers. Our science teacher was Mr Osbourne. We also had a teacher for German, Mr Herring. This was a new subject that we had just started to learn. Mr Light was our French master. We also studied woodwork with Mr Adams, and metalwork with Mr Gold. I do recall having to visit Mr Herring's home; he lived in Maida Vale, which was some considerable way from the school.

Our headmaster, Mr Pipe, was a gaunt figure. He had just returned to England from Burma, where he had been a Japanese prisoner of war. My recollection of him was of a very thin and emaciated man. But he took to his new job with enthusiasm. He had brought with him a large quantity of Japanese paraphernalia and memorabilia, which he had acquired towards the end of the war and which he placed on display for all the pupils to see and study. While at Choumert Road School I developed a liking for languages and won a prize for German—an inscribed German dictionary that I still have. I was also quite good at French. I also developed a keen interest in both woodwork and metalwork, no doubt related to my early efforts in Devon. Unfortunately there was no sixth form at this new school, so that at the age of 15 I had to leave. Mr Osbourne suggested I should stay on to take matriculation, but I would have been the only pupil there, so I declined the offer.

Printing

While I was still at school I was able to save up money acquired from a newspaper round, delivering newspapers and magazines. I used this money to buy an Adana printing machine at the Adana Works in Twickenham. It was a large metal hand press with a type area of 8 × 4 inches. I remember struggling home with this machine on underground trains and on buses. Once the machine was set up and working, I paid regular visits to a printer's suppliers located beneath the Holborn viaduct. On several occasions I ordered typesetting text, using the monotype process. The place I frequented was a typesetters on Red Lion Alley just off Fleet Street.

I also became interested in the related topic of lino cutting, a process that was done using mounted lino sheet. Using my home-made flat-bed machine, together with various type fonts that I had bought and also with lino cuts that I had made, I started to print a two-page comic called 'The Whizzer' in collaboration with my friend Albert Howes. Several editions of 'The Whizzer' were produced and distributed to my schoolfriends at tuppence a comic.

A further activity in which I was quite heavily involved was the making of lead toy soldiers. I had acquired a set of aluminium moulds of different soldiers, both modern and period. The soldiers were cast in lead and carefully painted in the various colours. A small concern opposite Belmont Buildings a few houses along dealt with the distribution of cardboard boxes of various sizes. I made enquiries there one day and was able to obtain a supply of long thin boxes, which I then used to package the toy soldiers in sets. I offered these to shops around Camberwell, and I was, in fact, able to sell some of these on a wholesale basis.

Yet another hobby in which I was very interested at the time was the making of fireworks. First of all I had the printing facility, so I could print coloured labels for the fireworks. But I had also acquired a textbook entitled *Spon's Workshop Receipts*, which gave some details about the preparation of various types of firework. After much experimental work, I was, in the end, able to make skyrockets, bangers, and various types of coloured and sparkling flare. As Guy Fawkes Day approached, I would spend many a happy hour making fireworks in order to mount a firework display for 5 November. Shortly after the end of the war fireworks were quite difficult to obtain. They had become really quite scarce, so the only way of celebrating Guy Fawkes Day was with my own home-made fireworks. I recall one incident with fireworks that almost ended in tragedy. I had made up an explosive mixture contained in a small cylindrical aluminium canister. These cylinders were ex-MOD supplies, used originally to hold five cigarettes for the armed services. I had added a very long delay fuse to the canister and foolishly thrown the device along the street, where it had rolled into the kerb. My ginger-haired chum from opposite Belmont Buildings, Kenneth Osbourne, picked up the cylinder, shook it, and examined it with curiosity. Concerned that it might explode at any minute, I screamed at him to drop the device. Fortunately he dropped the cylinder, and several seconds later it exploded, much to Ginger's horror and shock, and my relief.

La France, ou toute n'est que folie

Towards the end of my schooldays, when I was nearly 15 years old, Mr Light, the French master, encouraged one or two promising pupils in the class to take up penfriends in France. I was given the name of a family in France and started to write to one of the daughters of a Madame Cailly. We continued this correspondence for several months, and shortly after I left school I was invited to go on a visit to France. The Cailly family lived in a small village in the Somme region called Quend, Fort Mahon. I visited in the summer of 1950 and spent three weeks or so with the Cailly family. Unfortunately, on my arrival Monsieur Cailly had been taken very ill and subsequently died. Rather than send me home straight away, Madame Cailly arranged for her sister and brother-in-law to chaperone me and Madame Cailly's two daughters, Yveline and Micheline, for the following couple of weeks. It was in fact a very sad time for everyone, and a difficult time for me, because my French vocabulary was very much at the schoolboy level. I had very quickly to learn new words and new expressions in connection with the funeral, the church service, and so on. Unfortunately, apart from the two daughters, who had learned some English at school, nobody in the Cailly family could speak a word of English. But even Yveline and Micheline were really unable to discuss things or hold a conversation in English, so it was up to me to speak to all and sundry in French. This really was a baptism of fire for me, but at the end of my stay I was able to interact quite well, and my knowledge of French increased enormously.

In Britain we were still suffering post-war food rationing, and I was absolutely astonished to find that, although we had won the war, France at that time had no rationing. Sweets, chocolate, and most foodstuffs were not rationed. So while in France I took the opportunity shortly before returning to Britain to do some shopping. I remember buying a whole range of things, including sweets and chocolates, but also among the items of food stuffs I bought was a large smelly cheese, which I carried proudly with me all the way home.

Careers Adviser

A month or so before leaving school, all the boys were interviewed by a careers adviser with their respective parents present. In my case I was asked what I would like to do. I replied that I was quite interested in science. The adviser quickly pointed out that I had no qualifications and suggested that I consider something less ambitious. I then said that I had developed an interest in printing. I had a small printing machine, and I had also made a larger quarto flat-bed printing press. The adviser quickly settled on printing, and I was offered a job in the City of London at the firm of Ede and Fisher on Fenchurch Street. The job was an apprenticeship in the Bookbinding Department. I started work in January 1949 and quickly got the hang of bookbinding, but my heart was still in typesetting and in the composing room. In the course of my three or four months working at Ede and Fisher's I met another apprentice from the machine room, where the printing presses were. He said he knew a friend who was a compositor at another firm called Strakers on Commercial Way in Aldgate, who was willing to put in a good word for me. I went along to Strakers during one lunch break, and they had a vacant position, which I was offered. I started work as an apprentice compositor at Strakers around February 1949. I felt more relaxed in the new environment and I began to make rapid progress in typesetting in between other more menial tasks, such as making tea.

In the summer of 1949, I was nearly 16 years old and my parents, who had recently bought their first motor car, a brand-new Austin, took me on a one-week holiday break to Great Yarmouth (see Figure 11). While there, we visited Burgh Castle, some three miles inland from the coast, as well as other places of interest. Back in London, within a year or so, I was promoted to Assistant Ludlow machine operator. The Ludlow was a typesetting machine in which a line of type could be cast into a lead slug; this was used mainly for display headlines and especially for display advertisements.

Figure 11 With Mum and Dad in Yarmouth, 1949.

Evening Classes

While working at Strakers, I would walk along Commercial Street up to Whitechapel High Street and catch the bus home. The bus route went down Fenchurch Street to the Monument, across London Bridge, and then along the Borough High Street, Newington Causeway, to the Elephant and Castle. This took me to within a stone's throw of the Borough Polytechnic. I decided, therefore, to enrol in evening classes for five nights a week at the Borough Polytechnic, studying English, Maths, Physics, French, and German. Initially the idea was to matriculate, but, because of University of London examination changes, I subsequently took the General Certificate of Education. Since there were only five possible nights in the week, I was unable to take Chemistry at that point, but later on, when I took up a position at the Ministry of Supply in Westcott, Buckinghamshire, I was able to take Chemistry. My desire to study had been influenced by the son of a distant cousin of my mother, Lenny Clarke. He lived on Hopewell Street just around the corner from Picton Street. He had served in the army as an officer during the war but on demobilization had decided to gain university entrance and take a medical degree. Once he had acquired his degree, he applied for a hospital position in Kuala Lumpur and subsequently went there to live. While Lenny was still studying as an undergraduate he gave me one or two of his earlier books, which he had used when studying German for his matriculation. I still have these books today; one set of books was called *Deutches Leben*. He also gave me some thin books entitled *Fünf beruhmte Märchen*. I shall always be grateful to Lenny Clarke for the strong encouragement that he gave me at a very early stage in my academic career.

Also at this early stage, while I was still working at Strakers, I remember one morning reading the *Children's Mirror*. This was a section in the *Daily Mirror* allocated to children every Saturday. The piece that caught my eye was about an ex-Grammar School boy who had recently secured a job at the Rocket Propulsion Department (RPD) in Westcott, near Aylesbury. Because of my own growing interest in rocketry, I wrote to the editor of the *Children's Mirror*, explaining my interest. In the reply, I was advised to write to the Ministry of Supply enquiring about the possibility of my going to Westcott. After a week or so I received a reply inviting me to attend Adelphi House on the Thames Embankment for an interview. I attended the interview. They seemed quite impressed by my performance and suggested that I go to

Westcott, just a few miles beyond Aylesbury, for an interview at the RPD.

When I arrived at the RPD I was taken to the office to see a Miss Doney. She introduced me subsequently to two people from the Solid Propellant Division, Dr Crookes and Dr Errington. Following the interview I was offered a position as Scientific Assistant, on the understanding that I continued my education and obtained the O levels for which I was already preparing. It was 1952.

On returning to London, I resigned my job at Strakers and joined the RPD at Westcott. I took the necessary O levels during the summer of 1952, passed all five subjects, and settled down to working at RPD, in work that I was really enjoying.

Once I had joined RPD, and had a professional interest in rocketry, the desire to experiment at home waned, and my interests moved on to other things. There was just one small relapse later on, when I was teaching at the University of Nottingham. But more of that later (in Chapter 5).

The Dudman

I cannot remember exactly how I met Eddy Dudman. He is one of a group of friends that I had all living in or close by to Vicarage Road, Camberwell. Of the several friends in that area Eddy was a Grammar School boy and attended Archbishop Tennyson's School at the Oval, Kennington. Another two friends, Albert Howes and Charlie Lloyd, were also Grammar School boys attending Wilson's Grammar School next to St Giles' Church in Camberwell. On reflection, it seems that I alone among my schoolboy friends was a Secondary School boy. This fact did not seem to bother me or them.

Eddy lived in a largish house with a back garden, whereas I lived in a flat with no garden. Eddy lived with his parents and his younger brother, Kenny, but he also had a number of other friends, including Joe Ridge and Frank Butler. Joe, a Yorkshireman, was somewhat older than Eddy, having served in the Navy during the war, where he had been injured in an accident while wiring up a multiple rocket launcher. Joe's civilian job was in radio repair, but he was also a gifted pianist. At times all four of us would spend an evening talking or listening to Joe play the piano. Mrs Dudman was always most accommodating and rarely complained about the noise that we must have generated.

Because of my interest in rocketry and fireworks, I managed to interest the others in a number of hair-brained schemes that we tried out in Eddy's back garden. One episode I vividly remember was making explosive charges and using these to remove a large and unwanted bush in the garden. Another project was the generation of hydrogen gas, which was used to inflate a large balloon. One evening this balloon was released, carrying a large home-made thunder flash device with a slow-burning fuse. The idea was that it would ascend high in the air before exploding. Unfortunately, because of windy conditions and local vortices, the balloon did not rise much above the rooftops. But fortunately it drifted over towards a local recreation area, Brunswick Park, where both the thunder flash and the balloon exploded with an almighty bang.

On another occasion, we made an enormous gunpowder rocket and fired it in the back garden. It rose about 30 feet, turned to hit the ground, then slithered about before burning out. I remember taking an aluminium rocket round to Eddy's. It was based on the design used pre-war by the German *Gesellschaft für Raumschiffahrt*. The rocket stood about 4 feet high, but in the photograph was made to appear much larger. This device was never fired, but I did continue experiments to come up with a suitable fuel based on a mixture of nitro-cellulose, ammonium perchlorate, and magnesium powder, cast into solid blocks. Eddy always referred to these efforts as 'bang chemistry'. In fact he was quite interested in normal chemistry, and when he started work he joined an analytical chemical firm close to London Bridge.

At the age of 18 Eddy was conscripted into the Air Force. I changed my job to join the RPD and we lost touch for a couple of years. We linked up again after my National Service, but by then Eddy had married Pat, a fellow analytical chemist co-worker at his firm. I kept in touch with them both, on and off, for several years, until my wife, Jean, and I left England for the States. I vaguely remember that Eddy wanted a timer for his photographic work, so I offered to build it for him. The circuit was based on ex-War Department electronic components that I purchased at a number of shops on Shaftesbury Avenue in Soho. I built the timer and decided to publish the details in the magazine *Practical Wireless*, edited by F. J. Camm and known somewhat irreverently as Camm's Comic.

National Service

I assumed that because of the change of address from London to Aylesbury my call-up papers for National Service had got lost. Eventually, however, they caught up with me, and I was called up for National Service in November 1952. I applied immediately for deferment and attended a deferment board in Aylesbury, where a large lady, presumably a magistrate, was in charge of affairs. I explained to her that I was now working at the Rocket Propulsion Department at Westcott working on rockets. 'Rockets, rockets, what do you mean by rockets?' she roared. I tried to explain to her that had she lived in London she would have known exactly what a rocket was. But because she was in Aylesbury she had absolutely no idea or indeed no conception of the size and power of the rockets of which I spoke. She was clearly thinking about sky rockets for Guy Fawkes Night. It was later explained to me that, if I had had a university degree already, then I could have been deferred, but since I was at a very early stage in my studies it was not possible to have deferment.

My two years in military service started at Aldershot, where I spent around one month in initial training. I was later posted to Bulford Camp in Somerset. The training there was more specialized, and I was trained to be a storeman in the Royal Army Service Corps. Although I had originally wanted to join the Royal Air Force I was shoved into the army, so then I had requested to be in the Engineers or in something related to the work that I had been doing in civilian life. But my request was completely ignored, and I was forced into the Royal Army Service Corps as a storeman. I was, therefore, one of thousands of square pegs in round holes.

After several weeks of training, word went round that a new posting was imminent. As it happened, I had developed trouble with my teeth and had a large gum boil, which was the result of a wisdom tooth breaking through. I attended the Company dentist for treatment at roughly the same time as the posting was announced. The posting was for a whole group of Service Corps personnel to be posted to Egypt. As it transpired, this was the beginning of the Suez crisis. The whole group of newly qualified storemen packed up their kit bags and were posted. I was the only one left behind for continued dental treatment. I remember the dentist saying to me on my next visit that he would have to extract the wisdom tooth and that he had never undertaken such an operation before. He explained that it could be quite painful and very difficult for him. In fact the whole procedure went smoothly and within a day or so I was well on the way to recovery. Within the next week or so a new group of trainee storemen arrived and I was promoted to Lance Corporal to help train them.

While at Bulford Camp I went along to the Education Department and got into conversation with the Education Officer. I explained to him about my experience at RPD, and during the course of our conversation he suggested that it might be of interest if I were to give a talk on rockets. This I agreed to do. I went away to prepare my talk with diagrams and sketches. On the day of my talk there were quite a number of personnel in the audience, including one or two senior officers. I gave my talk, fielded a few questions, and after the meeting I was approached by a young private in the Royal Army Ordnance Corps (RAOC), a fellow called Roy Wallis. He told me that he had been called up at roughly the same time as I had. He also said that he had been truly interested in my talk and we continued chatting about rockets.

After that initial meeting we continued to meet whenever possible, and Roy, who was theoretically inclined, started to carry out various calculations on a number of aspects of rocket design, which we would discuss at subsequent meetings. This interaction continued for a month or so until I received word of my posting to Bicester. He said that he was very keen to work at RPD on his demobilization from the Army and we left it at that. I went to Bicester and wrote occasionally to Roy.

My posting was to the Base Ordnance Depot (BOD) at Bicester. There I joined the Command Ordnance Depot (COD), Royal Army Service Corps Group (RASC). Our major duties there were to supply rations to a whole group of army and air force bases in and around Bicester and the Oxford area.

Captain Carley-Pocock was the Officer Commanding and he had a young sprog Lieutenant under him. Carley-Pocock was a somewhat officious officer in charge and was very keen to make his mark with the military college and as a result everything was done by the book—that is to say, Queen's Regulations. I was ordered to take charge of the receipt of requests and subsequent supply of all food supplies to the surrounding areas in Bicester and other parts of Oxfordshire. This job involved receiving ration requests from various camps in and around the region. In addition, I was put in charge of company correspondence and had to learn to produce letters on a typewriter. One day I had a query with an order from the European Voluntary Worker Camp near Oxford. We referred to the camp workers as EV Wubbleyous. I picked up the telephone and dialled the camp number. Someone answered the phone and I asked to speak to the ration clerk. After a short pause the answer came: 'Russian clerk, Russian—I speak a Russian!' It took me some while to unravel this linguistic muddle. On occasions, when new deliveries of supplies arrived, I was called out to help unload trucks with food stuffs of various kinds. Once or twice I was also asked to help unload the meat wagon. That was extremely dangerous, and the quarter sides of beef were very difficult to carry. They weighed typically a couple of hundred pounds, almost twice my own body weight at the time. Carley-Pocock eventually left the unit. Whether or not that was anything to do with the fact that there was a major robbery from the station and several hundred weight of sugar went missing, I could not say. Of course, at the time, food rationing was still extant in the country. Sugar and other items of food that were stored in the depot were of great value on the black market. However, Carley-Pocock left, and a new Commanding Officer, Captain Smith, was appointed. Unlike Carley-Pocock, the new guy appeared a softer and more educated person. We got on well together, and I was quickly promoted to corporal.

With about six months of my two-year stint in National Service remaining, I was asked by the Commanding Officer if I would like to try for a short-term commission in the army. This meant being interviewed by the War Office Selection Board (WOSB). I agreed to have a go and was sent to the selection unit at Aldershot, I believe. There I joined a small group of nine other candidates. We stayed for about two or three days and were tested for a number of organizational skills and also public speaking. I decided to speak on the Mau Mau problem, a topical subject that was causing grief in Rhodesia at the time. This was a mistake, because I had to use notes. I realized at the time that I should have spoken about something with which I was completely conversant— namely rockets and the RPD at Westcott. I believe this was the reason that I failed to become an Officer Cadet. However, with hindsight, this was the best result for me at the time. I returned to Bicester to serve out my remaining time in National Service.

Towards the end of my second year at COD Bicester I made enquires about continuing my studies at the Oxford Polytechnic. I had noticed that there was a tilly that took people from the camp over to Oxford every week. I made enquiries about getting a regular lift over to Oxford so that I could take up my class. Course classes were held on one day a week, together with some evening work. I managed to get admission to the college and the new Officer Commanding was kind enough to allow me to take one day off a week to attend the college. In September 1954 I started the Advanced Level courses in Pure Mathematics, Applied Mathematics, and Physics and subsequently Chemistry at Ordinary level.

In November 1954, when I was finally demobilized from the army, I returned to the RPD at Westcott. I found that Roy Wallis, who had been demobbed a month or so earlier, was already at Westcott and working in the liquid propellant division. I saw him on occasions, but the very strong interaction that I had earlier experienced was now somewhat diminished. Nevertheless we did continue to interact and I learnt that he too was studying for university entrance.

Because I had already started the course at Oxford, I was able to get regular transportation every week from Westcott to Oxford. Of course on my return to Westcott I received considerable stimulus and support from members of staff of the Solid Propellant Division at the RPD. As well as Dr Crookes and Dr Errington, who were still running the Solid Propellant Division, there were several people there with whom I worked much more closely. These included Alan Shoulders, an Australian, Dr Ken Morris, a Naval Lieutenant on secondment, and Neville Morris, a chemist who had qualified from Salford Polytechnic just outside Manchester (see Figure 12).

My immediate supervisor was a Dr Lionel Dickinson, a red-haired chemist who was involved in research of new

Figure 12 The team at the RPD one year after I had left, 1957. Back row, left to right: Ray Heron, Lt Buchler, Ken Morris, Pappa Riedel, E. T. B. Smith, Nev Morris, Laurie Livemore, Eric Harrison, R. B. Davis, Eric Spalding, Robin Beeson, Bill Morrow, J. R. Smith. Front row, left to right: George Brock, Ben Barker, Jean Harbord, Dr Errington, Joan Roads, Dr Crook, Meg Bellingham, Lionel Dickinson, Col. Hardacre, Sqd Ldr Hermiston.

fuels for solid propellant rockets. Virtually all the solid propellant rockets that were being manufactured at Westcott at the time used extruded cordite, and the fellow in charge of solid propellants was a Colonel Hardacre. One day Colonel Hardacre received a letter in which the typewritten envelope had the letters 'cr' in his name accidentally transposed. He related this amusing story to us at the time, and unfortunately for him was for ever after known to some as Colonel Hardarce. One of the main tasks that I was involved in on the research side was the mixing and casting of new research propellants. This involved mixing materials together that were heavily doped with ammonium perchlorate, the oxidiser. The whole propellant consisted typically of a mixture of ammonium perchlorate and styrene monomer, together with a little accelerant to make the styrene polymerize into a solid block. The charges were then loaded into a small rocket motor, and I was invited along to see it tested. Typically the thrust and specific impulse of these various materials were measured to find the best combination giving the maximum performance. Fortunately for me I was very familiar with these parameters at the time as I had built a device at home for measuring and recording thrust and specific impulse of small rocket motors using an old gramophone turntable and a tin cylinder.

Another project with which I got heavily involved was connected with cordite propellants. Typically the cordite charge, which was usually shaped on the inside with a central conduit running the length of the charge, required an outer jacket of non-combustible material in order to inhibit burning on the outer surface. The outer casing was made of cellulose acetate in the form of a cylindrical sheath that was expanded and then heat shrunk over the cordite charge, together with solvent, in order tightly to encase the outer surface of the charge. However, it had been noticed on many occasions in test trials that there was a problem with complete bonding of the outer acetate casing to the cordite charge. It was, therefore, decided that an alternative technique for coating the charges should be experimented with, and I was asked to develop a winding machine that took a strip of cellulose acetate and simply wound it under tension around the charge. When the first coating was fully in position and firmly stuck down, a second strip of cellulose acetate was wound, starting with a displacement of half the tape width, so that it completely covered any gaps that may have occurred in the first winding. The basic cordite charges were made and extruded at a Royal Ordnance Factory (ROF). Some of the larger charges were extruded at the ROF at Renfrew in Glasgow, and at one point I was asked to accompany a critical piece of equipment that had been made at RPD Westcott for the extrusion of very large diameter charges up to 30 cm in diameter. The device that I accompanied from London to Scotland was called a 'Spider', and it was the device that made sure that when the charge was extruded the centre of the charge was missing. The shape and profile of the central shaft in the charge had an initial surface area that, as the charge burnt, maintained approximately the same constant surface area. The idea here was that a rocket when fired would produce a thrust that was constant over the burning period.

On another occasion I was sent to Aberporth on Cardigan Bay in Wales. There the Ministry of Supply had a rocket firing range, and experimental rockets were fired out over the sea. Up to that point I had witnessed static firing of rocket motors, both liquid fuel and solid fuel, but in Aberporth I was about to witness the firing of a live missile, and for me it was extremely exciting. I drew out my notepad and started to make a sketch of the launching ramp and the missile on the ramp. Shortly after I had started sketching in the hot sunshine I was aware of some shiny boots stamping by my side. I looked up and there was the Sergeant Major, who proceeded to arrest me and take me to the Security Officer. There I was questioned and when I explained that I was there on official business I was warned not to make more sketches and was released. However, when I returned to Westcott, I related the story to Lionel, who was extremely annoyed by the whole process. He picked up the phone and ordered half a dozen photographs of the missile and the firing of the missile, which he requested to be sent back immediately to Westcott and marked for my personal attention.

While at Westcott I obtained A levels in three subjects and also an O level Chemistry qualification. With these I applied shortly afterwards for university entrance. I was accepted for the special honours degree course in physics at Queen Mary College (QMC) in London. For the next nine months or so until I started at QMC in October 1956 I was free to relax from studies and enjoy the work with which I was increasingly involved at RPD. In order to supplement my income during this period I also managed to get a part-time job working evenings in Aylesbury. There I was involved in typesetting in a small jobbing printers for thirty shillings a week. Most weekends I would make the journey back to London and stay with my parents over the weekend, and it was on one occasion during the summer of 1956 that my parents and I were invited to the wedding of my cousin, Roy, son of my father's sister Molly and her husband, Jack. The wedding reception was held at Cronning Road home of the Delieu family and the bride Jeany Delieu. At this wedding reception I met a friend of Jeany and Roy's, a certain Jean Kibble, who lived a few hundred yards away on Commercial Way. For the next couple of months I managed some sporadic meetings with Jean while I was still working in Aylesbury. Then, at about the same time, I started my university degree course and began to see Jean on a regular basis.

3

Salad Days at University

Queen Mary College

After I had obtained my A levels, I wrote around to a number of universities and was accepted by two or three, but I chose Queen Mary College. Roy Wallis also chose Queen Mary College, but, whereas I was accepted for the special honours course in Physics, Roy was accepted in the Mathematics Department. We started college together, and, of course, I saw him quite regularly.

I started my university studies as a Fresher in early October 1956. There were initially thirty-one students in my year, most of them about 18 years old. I was one of the oldest in the year, but not the eldest. I found the first few months at college extremely difficult because of new and unfamiliar mathematics. For practical laboratory sessions we were asked to work in pairs, and my partner was David Conroy. He was one of the 18-year-olds and had come straight from school to university, but he seemed familiar with certain aspects of mathematics—for example, vector analysis, which he had been taught at school. The fact that there were areas and topics in the first year with which I was not completely familiar made me work especially hard, so that in the end-of-session examinations at the end of the first year, I was surprised—and I think one or two other people in the class were also surprised—that I ended up being top of the class. This result gave me great confidence to work even harder in the second year. At the end of that year there was no public announcement of the results, but I received a message to call on my second year tutor. He informed me that, if I continued to work hard, I could possibly achieve a first class honours degree in the final exam. It was at this stage that we all learnt that six or seven of our students had been downgraded to the ordinary degree while the remainder stayed in the class for a special honours degree.

The Head of Physics at Queen Mary College was Professor G. O. Jones, whom I had met initially at the entrance interview for the university. He had come from the Clarendon Laboratory in Oxford, and he had brought with him Dr Owen Davis as a Lecturer in Physics. We students thought of Davis as a brilliant theoretician, but I believe his original research work was in aspects of thermodynamics. The Reader in Experimental Physics was Dr Jack Powles. He had recently set up a nuclear magnetic resonance (NMR) group at Queen Mary College, and his principal interests at that time were in liquids. Another brilliant lecturer was Dr Heasty. He took Advanced Statistical Mechanics, while Dr Derek Martin took Advanced Electromagnetic Theory. Another lecturer, who acted as an adviser to the Royal Society and with whom I interacted much later on in my career, was Dr Roland Dobbs. He later became a professor, as did many of the lecturers at Queen Mary College, including Jack Powles.

So we moved into the second year (see Figures 13 and 14). It was around that time that Roy and I got together to discuss the idea of forming a new college society, the Interplanetary Society. After a suitable announcement, a meeting of interested folk was convened, and the society formed. I was elected President, Roy Vice-President, and Mr Baker, a physics student, was editor of our proposed journal, 'The Rocket'.

A cover for the journal was designed, and I made a two-colour linocut that was used by a local printer to produce a number of covers for the current and future copies of the journal. In fact, only two or three issues were produced in

Figure 13 Fellow second-year Physics students, 1958. Back row, left to right: Martin Sayers, David Worsfield, Norman Gilbert, Tony Clarke, David Horn, Philip Bradbury, Avril Kitson, Edward Papworth, Arthur Summerfield; middle row, left to right: Barry Blain, Roger Cullis, Frank Evans, Cyril, Sears, Peera Okera, Dick Baker, John Day, Jack Stone, Peter Mansfield, Paul Matthews; front row, left to right: Alan Fox, Ruth Sands, Cliff Sarfas, Ruth Fenn, Irfona Morgan.

the first year, with articles covering a range of topics. In my case I developed an interest in General Relativity and wrote an article on this for the journal. After that first year I renounced my Presidency to concentrate on my final undergraduate year.

During the third year, standard laboratory experiments were abandoned, and we all were allowed to choose a particular project set by the various staff members. I chose the project set by Jack Powles, which was to build an earth's field NMR apparatus. This had been done previously using valve technology; however, the requirement in my case was

Figure 14 Together with Dick Baker and Alan Fox in second-year Physics lab.

to design and build a portable apparatus using transistor electronics. This was 1958, and transistors had been around for only a very few years. The topic of transistor electronics had not been covered at all in the undergraduate course. In fact, it had been just two years earlier, in 1956, that the Nobel Prize for Physics had been awarded to W. B. Shockley, J. Bardeen, and W. H. Brattain for their researches on semiconductors and their discovery of the transistor effect.

So I had to read up about transistors and familiarize myself with what could and could not be done with them. One of the difficulties was that the transistors readily available were very low power OC 90s and OC 91s. This meant that for most of the electronics that I designed and built, I was obliged to use these very low power devices. The finished apparatus was taken out onto the front lawn of the college, by the clock tower, and a long coaxial cable was fed back to the physics research room, which was in the main building, just opposite the clock tower. Signals were obtained from water in a half-litre bottle. The bottle was moved around the lawn area, and the signal changed from a long ping to a short ping as we went over buried objects, pipes, and so on, that were located in, around, or under the front lawn. The basic signal received was at 2 kHz and could be heard clearly when amplified and listened to through earphones. We also ran a cable back to the lab, so that we could look at the signal on an oscilloscope. I believe it was a storage oscilloscope, so that it would be possible, if necessary, to photograph the signal at a later stage. My report on this project was written up and presented as part of my third-year practical work. At the time Jack was writing a short article on NMR for the as then new popular science journal *New Scientist*. He decided to include a brief mention of my project. The article was published in *New Scientist* and showed a photograph of me setting up and operating the equipment.

I really immersed myself in the third-year course and revelled in the interest and complexity of the various topics. We were allowed a certain choice in the areas that we studied in addition to a standard set of core topics, and I enjoyed every one of them. When it was time for the exams in the early summer I was very much on top of all my topics, and I passed all the exams with ease, ending up, as predicted, with a first class honours degree.

In earlier years I had taken various summer vacation jobs, including working at Eddy Dudman's firm. I used to work in the sugar assay department, testing sugar samples sent over from the docks by measuring the levulose content using the optical rotation of a polarized light beam. However, in my final year I decided to work for a company with which my brother Sid was employed at the time. The company was E. R. Watts and Son, just off Camberwell Road, who were scientific, mainly optical, instrument-makers—for example, they produced head up displays for aircraft. One day, while working in the research and development division of Watts, I was astounded to find Miriam Skinner in the office below the laboratory where I was working. She told me that her two brothers, John and Arthur, also worked at Watts, and of course I met them subsequently.

The research team I joined was led by a Dr Ian Young, a person I would meet again later in life and with whom I would interact quite strongly in work on magnetic resonance imaging (MRI). At the time he was in charge of the research project at Watts. I was asked to develop a piece of electronic circuitry to read and count the Moiré fringes in a circular diffraction grating. The object was to count the fringes created as the disc rotated and thereby measure small angular displacements of the disc. The project was based on valve electronics. The circuit comprised a series of Kip relays, Eccles–Jordan circuits, and Schmitt triggers. Today these circuits are called 'J K Flip Flops'. Various circuits were strung together to form reversible counters, so that the Moiré fringe patterns could be counted accurately for both forward and backward rotations of the disc. I found that, with a dozen or more valves in a circuit, it was very difficult to maintain the operational characteristics of the circuits, so that operational reliability was just not achievable. Of course, this unreliability is never mentioned in the textbooks, but in practice it was a real problem, and it was one of the reasons, I believe, why computers based on valve electronics never really took off in a big way.

Following graduation my interests and those of Roy Wallis began to diverge. He returned to the Rocket Propulsion Department, while I stayed on for my Ph.D. However, a year or so later I received an invitation from Roy to be best man at his wedding, which I readily accepted. This would have been around 1961. So my girlfriend Jean and I attended Roy's wedding to Mavis, but after that communication was reduced to the exchange of Christmas greetings. Even that declined when Roy and Mavis moved to Spain. We did hear from Roy shortly after the Nobel Prize announcement that he was back in England, but that sadly Mavis had died in 2001.

Jean was an extremely attractive young lady (see Figure 15). I remember her coming round to meet me at Watts one day, and jaws dropping and looks of envy on the face of all my male colleagues who happened to be around at that time.

Salzburg

While I was working at Watts that summer I met an Austrian student, Armin Karpf. He invited me and Jean to go to Austria to visit him and his family in Salzburg (see Figure 16). For transport I had an NSU Prima motor scooter, and the plan was to go on this scooter, with Jean riding pillion and

Figure 15 Passport photographs of me and Jean, 1958.

Figure 16 Jean (right) with Armin Karpf and a friend in Salzburg, 1958.

with as little luggage as possible carried in the pannier frame behind the pillion seat (see Figure 17). Armin's English was really quite good, but when we arrived in Austria and met his mother we had to communicate in German. Frau Karpf had a couple of spare rooms, and we were able to stay there for a day or two.

During the few days that we stayed in Salzburg we got to know a little of the town and the neighbourhood. One day Armin took us to the Salzkammergut, and there Jean and I went down into a salt mine via a sliding chute. We each sat on a mat provided and slid down around 30 metres. It really was quite exciting for all concerned. Later on we visited the famous water gardens at Heilbron, where concealed fountains rose through the seats around a large table (see Figure 18). We were regaled with stories of guests who had unwittingly sat on the seats and unexpectedly got soaked with water.

Figure 17 Jean with our NSU Prima scooter in France, 1958.

Figure 18 The Water Fountains at Heilbron, 1958.

We eventually bade farewell to Armin and his family, and set off for home, on our way passing through Vienna. It was while we were there that we observed the first Russian sputnik, which we saw moving slowly across the evening skyline. The occasion caused some stir on the streets in Vienna. It was early evening, and people stopped and looked up to the sky, pointing at the sputnik as it passed slowly overhead. It appeared to us as a very bright star moving across the evening sky.

On our way home we passed through France, and I stopped off at Quend, Fort Mahon, to show Jean where I had stayed with the Cailly family several years earlier. We saw Madame Cailly and one of her daughters, Yveline (see Figure 19); however, by then both sisters were married and living elsewhere. We stopped for an hour or so, then continued on our journey to catch the boat at Boulogne, eventually crossing the Channel to England and home.

Figure 19 Madame Cailly with her daughter Yveline and Jean at the Cailly residence in Quend, Fort Mahon, 1958.

Postgraduate Studies

Just before starting university I had been living in Buckinghamshire and I had applied for and been awarded a Major County Scholarship from the Buckinghamshire County Council. However, during my first term at Queen Mary College I was asked to go for an interview at the Ministry of Supply, and it awarded me a bursary that covered my university expenses from January of that first year and for the remainder of my time as an undergraduate. Also during this period I was promoted from Scientific Assistant to Assistant Experimental Officer. When I later obtained my degree I was to get further promotion.

Postgraduate Studies

I started my Ph.D. studies with some enthusiasm in October 1959. Also in the group was Iden Coleman, who was one of my contemporaries and had himself just graduated in physics. He later introduced me to dinghy sailing on the Welsh Harp, a small lake located just north of London.

I was placed in one corner of a large research laboratory. In another corner was Dan Pettie, the technician, and in a further corner stood a research rig used by Tony Hartland (see Figure 20). In the fourth corner was an area used by Dr Doug Cutler, a postdoctoral fellow who had stayed on for a year or so to continue his research.

Tony Hartland was working on liquid systems using pulse sequences to measure hyperfine interactions, and Iden joined with him to understand the rig and learn the nitty-gritty details of his experiments. The magnet that they had was similar to the one that was allocated to me. Both were made by Mullard, and they had magnetic fields of roughly 0.5 Tesla, making the proton resonance signals occur at 21.5 MHz. In my case I had the Mullard magnet but nothing else, and

I was given the task of building a pulsed spectrometer for studies in solids. This was to be the first work in the group connected with solids. Both Dan Pettie and Doug Cutler were very helpful in advising me and in some cases in helping me to pull together the necessary equipment to start my own experiments.

However, my primary task was to design and build a pulsed spectrometer, and I spent the first year of my Ph.D. studies designing and building a high-powered pulsed rig with the capability of producing a very short 90° pulse typically one microsecond in length. The electronics required to do this was really quite new to the group and no one in the group was able to assist me in the difficult task of designing the pulse amplifier and also the rapid response receiver system. I spent many hours in the library reading about radar systems and particularly some of the early radar designs

Figure 20 Tony Hartland demonstrating a Fourier transform to me, 1960.

where high-power RF transmitters operating at frequencies of around 5 MHz were pulsed on and off in a few microseconds or so. It was this approach that I finally adopted for the design of my own spectrometer. Of course, this approach implied that the final transmitter system that I built was incoherent. That meant that when one switched the pulsed oscillator off and then back on again, there was no guarantee that the phase of the RF pulse was the same as the previous pulse. For ordinary experiments, such as those I carried out in the study of certain polymers, this incoherence was not a problem.

In liquid studies it was common practice to follow the first 90° pulse by one or more additional pulses to produce a series of so-called spin echoes. However, in solid systems, such a sequence of pulses was not expected to produce a signal, because the relaxation time T2 was too short. However, one day I did try the application of a second 90° pulse some 6 μs following the first 90° pulse and to my surprise I occasionally found the appearance of what looked very much like a spin echo in a solid. As I watched on the oscilloscope, the echo jittered around, sometimes vanishing altogether and then sometimes restoring to a full echo. I mentioned the observations to members of the group as well as to Jack Powles. His initial reaction was one of disbelief. He thought the effect I had observed was a manifestation of pulse overload of the receiver system, an effect that causes initial blocking of the signal because of bias changes, followed by a slow recovery of the signal. I went away and spent another few days checking my results but remained convinced that what I had observed was a real effect of the spin system. I immediately set to work to try to understand this phenomenon and explain its appearance mathematically. I had read the paper by Lowe and Norberg on the theory and mathematics of free induction decays. I set to work, therefore, to extend that work in order to try to describe so-called solid echoes—that is, echoes in solids. Eventually I went to see Jack to present the experimental results, together with my theoretical ideas using the Dyson expansion. This is a mathematical series in increasing powers of time and the spin–spin interaction term. After I had discussed this with him, he suggested that it might be easier to look at systems with effectively two spins—that is to say, spin pairs. The material that he suggested was a single crystal of gypsum, hydrated calcium sulphate. This material had been used by Pake to observe in continuous wave experiments what was later referred to as the Pake doublet. A signal crystal of gypsum was obtained, and by then I had arranged to time the two pulses very accurately, so that the phase of the two pulses could be set and held at a fairly constant value. With this arrangement, the jittering that had occurred previously was no longer a big problem. Just as in the solid polymers, we found similar results in gypsum, and this work was written up and published in the form of a letter in the second volume of the new journal *Physics Letters*. However, in producing the experimental results I had noticed that there were some oddities associated with all these experiments, particularly if the second pulse was changed to a 180° pulse, but a proper explanation of this effect was never pursued and remains a mystery to this day.

During the second year of my postgraduate studies a new research student, John Strange, joined the research group. It was also around then that Iden Coleman left the group, and John took over the work on liquids while I continued with studies in solids. Jack's main interest in solids was in molecular motions, which occurred in these various polymer systems that I was studying. The results that I obtained at various temperatures were used to form a paper for publication at the Ampère Conference in Holland hosted by the Philips Company, where I made a short presentation entitled 'Relaxation Time Measurements in Various Polymer Systems'.

Salad Days at University 61

Apart from one British Radio Spectroscopy Group (BRSG) meeting in Bangor the previous year, the Ampère meeting was the first truly international meeting that I had been to, and I was greatly impressed by the lavishness of the reception laid on by the Philips organization. I remember, for example, the so-called Brabant Market area with stalls laid out with food and delicacies for all to sample and enjoy. In all there were about 500 people attending this conference and at the time it was the biggest gathering of people that I had witnessed for a conference meeting. The meeting was held in Eindhoven, and not only was I greatly impressed by Philips and its organization but I also found Holland itself to be an extremely agreeable, clean, and pleasant country. The conference was also an eye-opener for me, where I was able to see and meet the big names in NMR, people about whom I had read in the scientific papers and books. I came away from this meeting feeling that I had chosen well in my research topic and that I had really found my niche in NMR.

During the course of the third postgraduate year Jack asked me what I wanted to do following the Ph.D. He raised the possibility of going to the States for a postdoctoral period. Doug Cutler was also interested in going to the States, with his wife, Pat.

Jack mentioned one or two places that he thought might be able to accommodate me, but eventually suggested that I might fit in reasonably well with the group in Urbana, Illinois, led by Professor Slichter. So it was that I received a letter of invitation from Professor Charlie Slichter to go to Urbana, while Doug was invited to join a group in the Metallurgy Department, also in Urbana, led by Professor Ted Rowland.

❦ 4 ❦

Marriage and the American Dream

A Stormy Crossing

Jean and I were married on 1 September 1962 at St Giles Church in Camberwell (see Figure 21). We took our honeymoon in Scotland, staying at a small hotel in Inverness. We went by overnight train up to Inverness and a week later we returned, this time by coach, which was really quite uncomfortable. During the next week or two we lived at Crystal Palace, London, in rented accommodation while we were

Figure 21 Our wedding group, 1962. From left to right: Dad, Mum, Sid, Peter, Jean, Linda, Ma and Pa Kibble.

preparing ourselves for the trip to the United States. During this period I was offered some temporary postdoctoral teaching work at Queen Mary College.

We sailed to the States on the *Queen Mary*, and it took approximately five days to cross the Atlantic Ocean. The trip was fairly smooth most of the way but after about three days at sea we ran into a major storm, hurricane Ella, and because the ship did not have stabilizers it was tossed around on the high seas. On Gala Night, when all were expected for dinner, only a few of the passengers turned up to eat, including ourselves. Jean seemed to manage quite well and did not appear to suffer any ill effects, but I was feeling quite ill on that occasion. Nevertheless we attended dinner and managed to eat something. The storm cleared the following day. As we approached the American coast we heard on the radio news that a major Cuban crisis was brewing. President John Kennedy had warned Nikita Kruschev, the Russian Premier, that America would not tolerate the placing of Russian missiles on Cuban soil since they posed an immediate threat to the United States. He warned the Russians that, if they continued, drastic action would result. After disembarkation in New York, we caught a plane from Newark Airport down to Urbana, Illinois, and were met by Charlie Slichter and his wife, Nini, at the local airport in Urbana, Champaign. Charlie had arranged for us to be accommodated in an apartment at Orchard Downs, and Nini had organized a large food parcel to start us off in the new apartment, and the Welcome Wagon had brought one or two items for our use in the kitchen and so on.

Orchard Downs is about a mile or so from the Physics Department, which is on Green Street, in the middle of Urbana. When I reported to the Physics Department on my first day there was a state of pandemonium, with people practising air raid shelter drills and sirens being sounded from time to time because of the Cuban missile crisis. Within a week of my arrival I received call-up papers to join the US Forces. Charlie could hardly believe matters when I told him about the call-up papers and he immediately sent off a letter explaining the nature of my scientific work, thus effectively getting me excused from call-up. Within a week or so the Russian threat to the United States had subsided, when Nikita Kruschev turned back the boats that were about to deliver weapons to Cuba.

A little later that autumn, while we were still at Orchard Downs, I recall walking through Urbana one very clear morning and noticing on a large thermometer that was posted in the town centre that the temperature was minus 20° centigrade. I was walking briskly but as I continued to walk towards the campus my thighs began to get heavy and started to seize. I noticed that my eyelids crunched when I blinked and it was clear that the very cold weather was getting to me, even though the sky was clear blue and the sun was shining. I continued at a fast pace and eventually made it to the department. I was told later that it was really quite dangerous to be walking in that sort of weather, particularly as I had rather flimsy English summer clothes on at the time.

After a week or so we had settled into life in Urbana. Jean had managed to get herself a job at the University Health Centre, which was a short walk from the Physics Department. I had been walking into work but decided at some point that I really should obtain a vehicle. So I looked around and found a used car dealer in Champaign, who sold me a Ford Falcon Compact. I did not have a driving licence for a car, so I had to learn to drive rather quickly. A week or so later I presented myself for a test, passed, and obtained a licence to drive. In the meantime Jean had settled in well at the Health Centre and began to make many friends. Of course, she had to take an examination to qualify her for the job that she was doing, but she managed that quite easily and settled in to enjoy her time there.

Because of our long boat journey and other delays, when we arrived in Urbana, Doug Cutler and Pat were already there. They had flown to the States. In fact Doug was already installed with Professor Ted Rowland in the Metallurgy Department. He and Pat had also decided to live in an apartment in Urbana and were living not very far from our residence. We used to see them socially every other week or so and spent a fair amount of time in their company.

Fairly early on Doug had decided to acquire a car himself and bought a Triumph TR3 sports car. He and Pat would often go travelling in and around the Urbana area. At the end of their first year in the States, Doug and Pat decided to go for a long trip as a summer holiday. They travelled through a number of different States and on their way through Wyoming, at a small town called Medicine Bow, Pat was driving the car and took a bend too wide and in the process overturned the car. This was an open sports car, and unfortunately, as the car spun over, she must have hit her head on something, possibly the ground, and was killed outright. Doug, who was sitting next to her but strapped in, managed to clamber out of the car and, apart from a few minor cuts and bruises, had no injuries, although of course he was greatly shaken up. News of this accident came as a terrible shock and created a great pall over his stay in the States. It also had a big effect on us, creating a dark cloud that hung over the remainder of our stay in the States. About a week after the accident, Pat's body was recovered and taken to a funeral parlour in Chicago. We went with Doug to the funeral and the cremation service that took place in Chicago. The whole event was a very sad occasion indeed. After the funeral it took several months for Doug to come to terms with his loss, but he did stay in the post, working with Ted Rowland.

Sharp Cookies

Meanwhile I had joined Charlie Slichter's group. It was quite a large one and comprised a number of male members and one female, Judy Franz. I particularly remember Fred Lurey, the most senior person in the lab, who had served in the US Navy during the Korean conflict, Joe Asick, and David Ailion. While working in Charlie's laboratory, Fred Lurey finished off a few experiments connected with his work and spent quite a bit of the two-year period that I was there writing up his research thesis while I was modifying and building new equipment. When I first arrived in Professor Slichter's laboratory I got the distinct impression that Lurey was somewhat resentful, and he was not terribly helpful initially, as he seemed to take his seniority in the lab very seriously. But later, presumably when he was getting on with the writing of his thesis, he turned out to be very helpful, not only to me but also to other members of the group. However, once I had modified and built apparatus that was producing results, I had very little interaction with him.

David Ailion had only recently joined the group when I arrived. One of the problems that he had been set as part of his Ph.D. thesis was the study of spin locking in the rotating reference frame. I remember talking to him in some detail about the experiment that Charlie wanted him to consider. This was the measurement of rotating frame relaxation times. I recall spending some time explaining to him exactly what it was that Charlie Slichter wanted him to do. I also expounded that there could well be a difference of relaxation process in the rotating frame as opposed to the normal laboratory reference frame. During my discussions with David I must have said something to him that clarified things in some way, because thereafter he was always thanking me for helping him to understand the work that he was doing in rotating reference frames and even acknowledged discussions and conversations with me in published work later on.

There were, of course, a number of other people working in the laboratory, but I never really talked to them in any great detail about their work. One person who came as a visitor after about one year was Dr Dick Moran. He brought with him a great deal of NMR experience and very quickly became interested in some of the problems in which others were involved. His background knowledge in the theory of NMR was very impressive indeed and he was very familiar with all the mathematical techniques that were being used at the time—the density matrix and the Von Neumann equation together with all the detailed calculations that go into its solution. He took quite a broad interest in all that was going on in the group.

When I first arrived in the group, I was the only postdoctoral person present. Nevertheless I felt greatly inferior, because the range of knowledge that all the graduate students seemed to have of physics, electronics, and various other subjects appeared to me at the time greatly to exceed my own knowledge in these areas, particularly of theoretical physics. My feelings of inadequacy were not diminished by knowing that diagonally opposite my room was the office of Professor John Bardeen, a subsequent winner of a second Nobel Prize in 1973 together with Leon Neil Cooper and John Robert Schrieffer. The Nobel was awarded for their jointly developed theory of superconductivity. Just a bit further along the corridor were the offices of Professors Pines and Leon Kadenoff, who were also eminent theoretical physicists.

Charlie's interest at the time was mainly in the NMR of metals. He had read a paper by two research workers, Kohn and Vosko, who had published a mainly theoretical paper on the effect of doping in metals. They had performed calculations that predicted that, if one doped a material like copper with a small quantity of zinc, then conduction electrons in the copper would get scattered around the zinc centres. The new arrangement of electrons would fluctuate around the zinc centres, thereby changing the characteristics of the neighbouring copper atoms, causing slight chemical shift changes that varied with decreasing amplitude as the distance from the zinc scattering centres increased. This meant that one should see slight shifts in resonance frequency of the copper atoms surrounding the zinc centre. These small resonance shifts could be detected by performing a double resonance experiment between the Copper 63 and the Copper 65 resonances. My task, therefore, was to build a double resonance spectrometer with the capability of detecting the slight resonance shifts, but in addition I also had to create the doped copper crystals.

My initial challenge was to draw from the melt single crystals of doped copper, the melt comprising pure copper doped with one or two per cent of zinc. Once the single crystal of doped copper was produced, the next process was to slice the copper sample into a number of thin slivers, about ten in all, in order to increase the surface area to produce a larger NMR signal.

The slicing process was carried out using a spark cutter. This process completely upset the smoothness of the resulting slice surfaces, so that it was then necessary to etch away the material back to the smooth and undisturbed shiny copper surface. The ten or so slices were then carefully reassembled with suitable non-resonant spacers between the slices so as to re-form the complete copper sample. The double resonance spectrometer and the sample between them took about one year to produce. After many attempts to see the predicted perturbations of copper resonances from normal resonance due to doping, I concluded, reluctantly, that they were either too small to be seen with the equipment I had constructed, or not there at all.

During my period in Urbana I received a pre-print from Jack Powles of a paper that he had co-authored with John Strange. This was a general paper entitled 'Solid Echoes in Systems of Spin One Half'. Rather than use the Dyson series expansion suggested in my Ph.D. thesis, Jack and John had used a similar time series expansion based on the second and higher even absorption line shape moments to describe the free induction decay and the production of an echo. By this means they were able to describe all the results that I had obtained originally and John had obtained subsequently on polymer materials. Although Jack and John's paper was a series expansion approach, it opened up the field in solid echoes, and immediately I started to think about an extension of the theory in general to spin systems greater than spin one half. I also extended the theory to the study and production of solid echo type signals in two spin systems and in perturbed systems with spin greater than one half. Towards the end of our second year in Urbana, about a month or two before we started to pack up to return back to England, I managed to get Doug to do a few experiments for me on work that I had been doing as a sideline to my main task while in Professor Slichter's group. I was particularly keen to do one or two additional multiple-pulse experiments. Basically, they were experiments in which we looked at solid echoes, in the case of Doug Cutler's work, in aluminium powder. I was especially interested in extending the solid echo theory that had been developed by Powles and Strange, to consider the case of materials with spin $I = 5/2$. At the same time I was also interested in looking at two spin species systems, and in correspondence with John Strange he agreed to look at the fluorine resonance in a sample of sodium fluoride. But, because of the presence of a second abundant spin species, sodium, with spin $I = 3/2$, we were able to see novel solid echo effects in one and two pulse sequences. In the paper that I was writing up at the time I got permission to

Marriage and the American Dream 71

include the very preliminary results that had been obtained both by John Strange and by Doug Cutler. This paper was accepted by *Physical Review* and published in 1965.[†]

Unfortunately during the period that I spent in Urbana it was not possible to pursue multiple-pulse techniques. The facilities within Professor Slichter's group were really not designed for short pulse applications in solids. Furthering the experimental side of solid echo work had to wait until I returned to England in September 1964.

During the two years that we lived in the States, Jean and I took two long summer vacation trips of about four weeks each. I had had no particular problems with the car that I had bought earlier. So on our first summer vacation visit in the first year that we were there, we did a grand tour of some of the Southern states. We left Urbana and dropped due south to Perduca, over to Cairo in Illinois, and then further south to Memphis in Missouri, then further south to Jackson, Missouri, and even further south to Lake Pontchartrain, close to New Orleans. We arrived at a campsite at about 7 p.m. It was quite dark, but I was able to erect our tent in the headlight beam of the car. I noticed hundreds of flies and mosquitos buzzing around the beam. We erected the tent and netting and I felt I had been bitten around the ankles, despite the fact that I was wearing socks. The following morning both my ankles were severely swollen and I needed medical attention. On striking tent we made for New Orleans to find a hospital. We eventually turned up at the hospital emergency wing, where I was examined by a doctor. He prescribed some antihistamine tablets that helped to reduce the swelling. While we

---

[†] In November 2011 I received an e-mail from a Dr Fradin, claiming that it was he and not Doug Cutler who had obtained the results in aluminium. In his note he says that at the time he was a research student in Ted Rowlands research group, and furthermore he had constructed the NMR equipment used to obtain the results.

were waiting to see the doctor, I noticed large signs in the waiting room saying 'Whites Only'. From New Orleans we went south through Baton Rouge, straight along the Old Spanish Trail to Houston in Texas, then continuing on to San Antonio and further west to El Paso, crossing the Rio Grand, a dried-up trickle, and into Juarez on the Mexican border before returning to El Paso. We continued on the Old Spanish Trail to Tucson and Phoenix in Arizona, where we visited an old abandoned film studios. After settling in our hotel in Phoenix we drove out to the abandoned studios.

By then dusk was approaching and as we drove along the road to the studios my headlights caught what looked like a huge spider in the road. I stopped the car and got out and as I approached it I realized that there was not just one spider; there were dozens of these huge spiders creeping along from the sandy sides of the road. I took a stick from the ground and tried to prod this large spider, which was in our way. Initially it would not move, but eventually I managed to clear it and other spiders away. We wanted to continue but by then it was really quite dark and we could see little of the film set itself. So I attempted a three-point turn on this narrow road. Unfortunately the back wheels of the car slipped off the tarmac into the soft sandy verge and started to spin, sinking further into the sand. As Jean could not drive at the time, I suggested that she get out of the car to help rock the vehicle out of the ever-deepening sandy furrow. The spiders were still around but had retreated somewhat. With only flat shoes and summer shorts, Jean finally got out of the car and pushed. After a couple of attempts the car was rocked back up onto the tarmac and Jean was back in the car shaken but unharmed. I later learnt that these spiders were tarantulas and that they could spring 6 feet or more to catch their prey and quite often caught birds that were unawares. We then returned to Pheonix, and the following day we made our way up to Flagstaff then down and across by the Hoover Dam to Las Vegas. From Las

Marriage and the American Dream 73

Vegas we started to move northwards into Utah through the Zion National Park and up to Salt Lake City. From Salt Lake City we drove east over to Denver in Colorado and then continued eastwards first to Kansas City and then on to Colombia and Missouri, and then, after visiting Springfield, Illinois, the birthplace of Abraham Lincoln, on to Dekatur. With a mixture of camping and stopping in small hotels, we continued our round trip and ended back in Illinois.

In our second trip we went due south from Urbana down to Nashville, Tennessee, and then over to Atlanta, in Georgia, dropping further south on to Route 75, which took us pretty well down through the centre of Florida to end up in Miami. In Miami we parked the car and caught a boat that carried us over to the Island of Nassau, where we stayed for a week before returning to Miami and making our way back home to Urbana.

Home on the Range

In Urbana itself we were made very welcome. I remember on one occasion being invited to a farm by one of Jean's friends, and taking a hayride around the cornfields. The corn, of course, was corn on the cob, and it really was 'as high as an elephant's eye', as it says in the song from the musical *Oklahoma*. Our trip ended with a barbecue and huge T-bone steaks, each one pound in weight, were cooked on an open barbecue. I had never seen so much steak served up in one helping. We all enjoyed the barbecue, with fresh home grown tomatoes and other vegetables. The whole event was a memorable occasion. We were invited to celebrate Thanksgiving at the Slichters, where we met the whole family, including Charlie's sons. On another occasion we were invited back to the Rudecil family for a meal. All in all we were made to feel very comfortable with the very friendly group of people that we got to know while we were living there. Of course, it made our parting that much more difficult when the time came to leave.

During the summer of 1964 I received a letter from Raymond Andrew, who had been my Ph.D. examiner. He had recently moved from his position in Bangor to take up the Chair and Head of the Physics Department in Nottingham. That summer Jack Powles came over to the States and visited us as well as an old friend of his in Illinois, the astrophysicist, Professor G. C. McVittie. I had been awarded a prize for obtaining a first class honours degree at Queen Mary College and by pure coincidence I had chosen the book by McVittie entitled *General Relativity and Cosmology* as the Department of Physics prize. I was, therefore, amazed to find that Professor McVittie was right there in Urbana during the period that I was there. I suspect he had actually emigrated from Britain and had settled there on a permanent basis.

While Jack was visiting Urbana, he mentioned to me that he had recently been appointed to the Foundation Chair of Physics at the new University of Kent at Canterbury and he was looking for staff members to join him. I told him that I had already been offered a place at Nottingham and I was about to go to New York for interviews with the government recruiters to a whole range of research establishments in Britain. Subsequently, we went to New York and as well as the option to return to the Rocket Propulsion Department at Westcott, I was offered a general position in the Scientific Civil Service as Senior Scientific Officer but not assigned to a specific centre. Among the available jobs there was one at the National Physical Laboratory (NPL), and when finally we returned to England I went to NPL to consider the possibility of joining them. As it happened Tony Hartland had returned a year earlier and he had taken up a post there.

One year after I had left Queen Mary College, John Strange finished his Ph.D. work at Queen Mary College and went to

Marriage and the American Dream 75

the United States working as in a postdoctoral capacity at Cornell University, Ithica. He decided to stay in America for just one year and returned to England pretty well at the same time as Jean and I returned.

We returned by ship on the USS *United States of America*. Unlike our crossing to the States two years earlier the return was uneventful, and we arrived back in London five days later. We had crated up most of the heavy luggage and material that we had acquired during our two-year stay in America. That eventually arrived a week or so later and was then directed on to Nottingham. The date was now early September 1964 and we spent a week or so in London visiting our friends and relations, each of us staying with our parents, until eventually we made our move to Nottingham.

5

The Wanderers Return

A Job with Prospects

We arrived in Nottingham in late September 1964. We had travelled by train and turned up at the campus to meet Professor Andrew and other staff members. We were accommodated for about a week in the University Staff Club, which had a couple of bedrooms for visiting guests. During that first week we looked around for properties and combed the newspapers for accommodation. Eventually we found a vacant property in Beeston adjacent to the campus. It was a small two-bedroom house, and we decided to rent it for a short period.

In the meantime we had noticed a building site in Chilwell, where numbers of new houses were being constructed. We made enquiries and found that there were several possibilities for buying a new property, but any option would take at least six months to complete. The property that we settled on was No. 7 Clarke's Lane, Chilwell. When we viewed the site, the house had just had footings installed and the main structure was about one foot high above ground level. The agreed purchase price for the property, including one or two extras that we required, was £4,800. My starting salary at the university was just under £2,000 per annum, but we had managed to save money during our stay in the States and were able to make a substantial down payment, so that we had no problems borrowing the difference as a mortgage.

The house was almost opposite the old Chilwell Manor. But, as building proceeded, we were dismayed to find out that the Manor House was scheduled for demolition. In its place and on its land were built subsequently a dozen or so modern three- and four-bedroom houses, thereby destroying forever the character of this once charming estate.

Our house was duly completed on time and we were able to move from our rented accommodation into our new house in March 1965. Jean took full-time employment in an office at the Raleigh works on Middleton Boulevard in Nottingham.

I had started lecturing and at one point I had the responsibility of the second-year undergraduate laboratory. Here I was responsible for introducing one or two bench-top experiments on aspects of aerodynamics. These tests were directed to measure lift and drag on model wing sections. I later built a much larger wing section at home and used this to measure lift and drag. Later still, as part of a third-year project I built a large model helicopter with a twenty-four-feet diameter rotor blade. The whole device was pedal powered (see Figure 22). I got students to pedal and measure the lift as part of a third-year project. With the three feet chord width of the rotor blades, the maximum lift measured with me pedalling was about 80 lb—not enough to lift off—alas!

Multi-Pulse NMR

As part of my non-teaching activities I was also setting up a laboratory with a new Research Student called Donald Ware, who had joined me from Canada. He had come from the University of British Colombia and had worked in the Chemistry Department for his M.Sc. under his supervisor Professor Basil Dunnell. I explained to Don that I wanted to build a new multi-pulse spectrometer capable of producing trains of pulses. He quickly became immersed in this, although he had very little experience of electronics up to that point. Nevertheless, within a year or so we had designed

Figure 22 Testing the pedal-powered model helicopter in the Sports Centre, University of Nottingham, 1970s.

and built a spectrometer and started to use it to look at the NMR response in a range of solids. I had acquired a starter grant from the Science Research Council (SRC) and was able to purchase a Varian electromagnet and power supply. This could be easily tuned so that we could see signals, not only from proton samples but also from fluorinated samples. Very soon afterwards we tried the multiple pulse experiments, which the machine had been designed to observe. The sample I chose was a single crystal of calcium fluoride. That was mounted on a rotatable sample holder so that we could observe the NMR signal when the crystal was orientated with the magnet axis pointing along either the [111] or the [100] crystal direction. When we applied a string of coherent 90° pulses to the sample with a suitable phase shift between the first pulse and the string of pulses, we found to our astonishment that we could maintain the FID for very long times. In fact we looked at different pulse spacings and at one point we were able to observe the free induction decay for over 1 second even though the T2 of calcium fluoride in the [100] direction was about 40 microseconds. Of course to do this experiment we needed short 90° pulses and these were one microsecond long with a very fast recovery, allowing us to observe the nuclear signal between the pulses.

All these results were obtained early in 1966. A few weeks later Jack Powles was invited to Nottingham to give a talk at a departmental colloquium. After the talk he was shown around the various NMR groups and eventually came to my laboratory, where I showed him our latest results on calcium fluoride. His initial reaction was that he thought he had seen something like this previously but he could not remember the detail. He had received a pre-print that he thought was from John Waugh at the Massachusetts Institute of Technology (MIT), but he would check this and let me know subsequently. A week or so after he returned to Canterbury he sent me a copy of a pre-print that he had received from John Waugh that also described multiple pulse experiments in materials.

In the intervening period, while I was waiting to hear from Jack Powles, I wrote up our material in the form of a letter that was sent off for publication. Our paper was published subsequently in the second volume of the new journal *Physics Letters*, a week or so after John Waugh's paper was published in *Physical Review Letters*. This was 1966. Waugh claimed later that he was unaware of the previous work that had been done in London by myself on the two- pulse experiments in Gypsum published in 1962 or the subsequent work of Powles and Strange published in 1963 or indeed the further work that I had published in *Physical Review* in 1965, and seemed determined to claim that he and he alone had thought of doing the multiple pulse experiments in solids. This attitude led to a great deal of bad feeling and internecine squabbling, which started on the publication of our paper in *Physics Letters* and continued through the years until the early 1970s.

It was also in 1966, in the summer, that my father died of a sudden and unexpected heart attack. He was 61 years old.

In 1967, our first daughter, Sarah Jane, was born. Three years later we were blessed with our second daughter, Gillian Samantha. Whereas Sarah's birth was quite normal, Gillian's birth was a breech, which led unfortunately to complications with one of her hips. This necessitated the use of a correction harness for the first six months or so, until the problem resolved itself.

The Research Continues

From 1966 to 1972 I had a series of very bright and able students and postdocs who joined me in our work on multiple pulse NMR. Among these were Ken Richards, David Needham, Dennis Stalker, Peter Grannell, and a postdoc Allen Garroway, an American, who had done his Ph.D. studies in the Physics Department at Cornell University. During the same period John Waugh had assembled a formidable group of research students and postdocs also working on various aspects of multiple pulse NMR. Among these were Alex Pines, Ulrich Häberlen, Michael Mehring, and Bob Griffin. In order to understand and develop the multiple pulse sequences, John Waugh had used the mathematical techniques based on the Magnus expansion and established the idea of an average Hamiltonian over the pulse cycle.

In Nottingham I had proposed the idea of the logarithmic operator to describe the evolution of multiple pulse spin echo sequences in solids. Using this I was able to reproduce all the results that Waugh and his group had produced using the Magnus expansion, but in addition the logarithmic operator approach had the flexibility that enabled me to introduce the idea of reflection symmetry into the calculations. As a result of this theoretical approach I was able to design new sequences based on reflection symmetry. These sequences offered the possibility of removing higher orders of the remaining dipole–dipole interactions. The result was that much greater line narrowing could be produced than could be achieved with the four pulse cycle that John Waugh and Ulrich Häberlen had devised. I was able to extend further the reflection symmetry approach. By this means I was able to remove even higher orders in the logarithmic operator expansion so that, in principle, the next highest order of remaining interaction was eighth rather than sixth.

In the race with John Waugh to explain the way in which the multiple pulse sequences varied as a function of the pulse spacing, I suggested initially the idea of projecting the loss in echo amplitude at the first echo. This led to an exponential loss in signal amplitude. The time constant for this signal loss process was theoretically proportional to τ^3 where 2τ is the time between successive 90° pulses. We rushed into print with this, but it turned out, on closer examination, that the signal decayed with a time constant proportional to τ^{-5} rather than τ^{-3} as we had expected originally. So the next idea was to project the signal loss of the second echo rather than the first echo. This meant working out the actual details of the solid echo after two pulses and taking that loss in signal from the original normalized amplitude and projecting that loss. This time the time constant for the echo train observed experimentally and the theory agreed very well. We tried the theory in calcium fluoride and that was fine; we also tried it in a phosphorus compound, possibly phosphorus pentoxide, looking at the phosphorus resonance, and again the theory and the experimental results seemed to agree rather well. But of course what we were really interested in doing was developing a theoretical explanation and a technique that would remove the effective dipole–dipole interaction in solids while leaving bare for observation any underlying spin interactions. For many of the materials at which we were looking at the time, the underlying interaction was the chemical shift.

At this point I deviated somewhat from multi-pulse sequences to look at various spin locking sequences, because we had observed an interesting effect for which, initially, we did not have an explanation. That was that, when one spin locked or if one executed a closely spaced multi-pulse sequence, it was observed that the amplitude of the signal appearing between the pulses seemed to have a slight slow oscillation. When we did the actual spin locking experiments, we could see this very easily. By plotting out the spin lock signal amplitude as a function of the spin locking pulse length, we were able to trace out this variation in a number of situations and also at a number of spin locking B_1 fields.

To return to the spin locking experiments, it turned out that, if the spin locking sequence was left on long enough, the initial transient oscillations that we were seeing died away and produced a spin locked signal amplitude that could be calculated from the thermodynamics of the system. This agreed very well with the work that had already been done by Charlie Slichter and by Irwin Hahn and Sven Hartman, who had originally produced the idea of spin locking and had applied the concept to double resonance in the doubly rotating reference frame. The transient oscillating signals that we had measured had not been discussed or mentioned previously, so it was, in my view, a rather nice piece of work that allowed us to study a spin system as it approached a final state described by a spin temperature in the rotating frame. We started off with a non-spin-temperature and a non-equilibrium situation. It took a few wiggles or oscillations before the system settled down to a steady value and could be predicted purely from thermodynamics.

Although this work arose from an initial interest in multiple pulse sequences, it was really a separate piece of work that stood on its own and was very satisfying. For this work and the multiple pulse work, in order to calculate the magnetic resonance details, I adapted and developed the cumulent approximation technique that allowed us to go to somewhat higher orders of spin–spin interaction in calculating the residual error terms. The cumulent approach was an approximation, but, as it turned out, it was quite successful, particularly for the higher order interactions, and allowed us to calculate sixth-order and eighth-order moments of the absorption line shape in terms of products of lower order moments that could be calculated exactly. This approach was extremely useful in some of the rather heavy and lengthy calculations that were involved in this and other work that was proceeding at the time.

Interesting as it was, this deviated from the multi-pulse work and allowed Waugh's group to make progress. They eventually came up with the so-called WaHuHa sequence, and this was really quite successful in removing the dipole–dipole interaction, but at the same time leaving behind a residual chemical shift interaction. I say residual interaction. It was actually a scaled interaction because what happened is that the residual interaction got reduced somewhat by a scaling factor. This was calculated for that pulse sequence and for other pulse sequences. It was at this point that we returned to the question of chemical shift and the study thereof and started again in earnest to compete very strongly with John Waugh.

In the second round of this interaction and competition with John Waugh, as explained earlier, I proposed the symmetrized pulse sequence and further applications of the symmetrized pulse sequence that I have already discussed.

Peter Morris, who had just recently joined my research group as a Ph.D. student from Cambridge, tried several of these more complicated reflection symmetry cycles and found that their line narrowing efficacy was difficult to maintain and the stability of the cycle was not as good as it appeared to be theoretically, so he settled for an intermediate cycle, which had quite good line narrowing qualities. However, the more complicated eighteen or more pulse cycles I had proposed and published earlier in 1971 used full reflection symmetry and were later applied by Burum and Rhim in Pasadena. They were able to produce outstanding line narrowing for the proton resonances in a single crystal of hydrated calcium sulphate. This work was published in 1979. It was around this point that the Waugh group's interest in extending multiple pulse techniques waned somewhat. Their collective interest seemed to turn to questions relating to the general problem of reversibility in multiple pulse decays, and they devised a so-called magic pulse sequence, which has the ability completely to reverse a free induction decay, which had, to all intents and purposes, completely decayed but then suddenly was made to start growing, reappearing at a later stage as a free induction growth signal. My own interest also seemed to wane somewhat in multiple pulse techniques and adaptations of multi-pulse techniques.

In the early 1970s Ulrich Häberlen returned to Germany from Waugh's group at MIT and took a position at the Max Planck Institut für Medicinische Forschung in Heidelberg, originally the Kaiser Wilhelm Institut, but renamed, in the 1930s (I believe), in honour of the eminent physicist Max Planck. Another German, Michael Mehring, returned to Germany from Waugh's group and took up a post at the Technical University at Bochum in Germany.

I first visited Heidelberg at the invitation of Professor Karl Hauser. This was to give a talk on my work on multiple pulse methods in solids. There were several questions from the audience, but Hauser remarked later that my trousers looked as though they would slip down at any moment. The talk was well received, despite Hauser's remark, and I was later invited to return to Heidelberg to spend a sabbatical year working with Ulrich Häberlen and colleagues, which I accepted.

Allen Garroway had joined my group the previous October, in 1971, from Cornell University, where he had completed his Ph.D. studies with Bob Cotts. When he and his wife, Mary, a Japanese American, arrived with their backpacks, they looked as though they had been hiking from London Airport. In the autumn of 1971 we had been suffering a series of electricity power cuts, and on the day that the Garroways arrived in Nottingham we were in the middle of an electricity cut. They had been invited round for a meal in the evening. Fortunately we had a gas cooker and could prepare and cook the meal. However, when they arrived all that we could offer them was a dimly lit dining room, using an old oil lamp and a candle-lit supper.

Earlier in 1971 I had received a major SRC grant to computerize the new spectrometer that we had constructed. In the summer of 1971 I was able to purchase a Honeywell desktop computer with 4K of magnetic memory. I had started immediately to wire in changes so that the computer could be used to control the RF pulses and gating on the spectrometer to produce a series of pulse sequences for our experiments. This was operational when Allen Garroway arrived, but he threw himself into the next major task, which was to implement the fast Fourier transform algorithm of Cooley and Tukey, and this had to be done in the 4K of available memory. By early 1972 the machine was up and running and performing fast Fourier transforms. This allowed Allen, Peter Grannell, Peter Morris, and others to proceed more rapidly with their experimental work.

In early 1972 the main thrust of our work was the development of multiple pulse techniques and their applications to study chemical shift and chemical shift anisotropies in a range of compounds, particularly fluorinated compounds that we were studying at the time. By Easter time of 1972 we had exhausted all the fluorinated compounds that we had, including any that we were able to acquire from the Chemistry Department.

In those days in was customary for all Physics staff and research students to take their morning coffee in a common room, or tea room, within the Physics Department. Refreshments were served there from 10.45 to 11.15 each day. In early May, I believe, Allen Garroway, Peter Grannell, and I went along to have our coffee at about eleven o'clock. We were bemoaning the fact that we had used all our samples of fluorinated compounds and wondered what to do next. During our discussions the rest of the tea room members had slipped away, and by 11.30 that morning the three of us were left still busily discussing what we might do and what our future plans should be with the new computerized spectrometer.

I had been grappling earlier with a paper by Kubo and Tomita, who had written a theoretical paper on the NMR of ferromagnetic systems. Their paper had been written entirely in terms of k-space, rather than real space (r-space), which was customary for solid state theoretical calculations. It was while reading this work that I began to have the idea that one could in principle look at the spatial structure of materials by observing the signal response in k-space. Of course, in ordinary NMR the free induction decay is a function of both time, t, and the angular frequency shift, ω, describing the absorption line shape. These are conjugate variables, so that the absorption line shape and the free induction decay are related by the Fourier transform. However, if the free induction decay is observed during the application of a magnetic field gradient, the absorption line and the free induction decay are then related by the new conjugate variables r and k, where now k is a function of both time, t, and the gradient strength G, and where r is a spatial displacement.

All these things were passing through my mind during our discussion in the tea room that day, and as a result of these thoughts I suggested that it might be possible to carry out experiments using a magnetic field gradient in combination with the line narrowed experiments that we had already achieved. If we could apply a sufficiently large gradient, I surmised, we ought to be able to see the actual atomic structure in these materials, at least theoretically, in what amounted to an NMR diffraction experiment.

Allen Garroway, who had worked with Bob Cotts using magnetic field gradients, was rather skeptical initially. Peter Grannell, who was in his third and final year of Ph.D. studies, was more receptive to the idea. Following our discussions, I went away to write down my ideas in more specific form and produced a wadge of calculations that I subsequently gave to Peter for his comments. Because Peter was finishing his Ph.D. studies, there was very little that he could do to make headway on these new ideas. We did discuss the possibility of using some old equipment built by a former student, Alwyn Finney. He had built a gradient coil to do work on signal storage in magnetically broaden systems. This coil was tried out, but, in the event, Peter decided finally to build his own gradient coil system. Construction of the new coil was undertaken later in 1972, by which time I had departed for Germany for my sabbatical stay in Heidelberg. I do recall, however, that, before departing, I made the comment to Peter that I thought that the NMR diffraction idea could one day lead to a Nobel Prize.

6

Alt-Heidelberg

> du feine du Stadt an Ehren reich,
> am Neckar und am Rheine
> kein andre kommt der gleich.[†]
>
> Scheffel

Die Einladung

In 1972 we were still living at 7 Clarke's Lane, Chilwell, and I had already made arrangements for a visiting professor to rent our house while we were away on sabbatical. We drove down to Dover and got a mid-afternoon ferry across to France. Then I drove through France and partway on our way to Brussels we stopped at a small town called Ath. By the time we reached Ath the evening was drawing in and with two children on board feeling very tired and sleepy we decided to stop in the middle of the town. I found a parking lot, simply wound the car windows up and got comfortable in the car for a night's stay in Ath. At first light the next day we were packed up and ready to go, and we made our way towards Brussels. We entered the outskirts of Brussels at around seven o'clock in the morning, and the streets were deserted. As we moved into the centre, we came across a traders' market. People in the market area were loading fruit and vegetables onto their wagons. A little further on we were

---

[†] *Old Heidelberg, thou fine city of a noble land, on the Neckar and the Rhein, no other comes its equal.*

able to find somewhere, a stall I believe it was, where one could buy a cup of chocolate or coffee and something to eat. It was a makeshift breakfast, which we all enjoyed, and then we continued our journey through Brussels. Because there was really no traffic about at that time, we managed to pass right through the city centre out the other side of Brussels and towards the German border. We finally got to Aachen on the German border, and there we were able to pick up the autobahn, which took us all the way to Heidelberg.

With stops for lunch and refreshments, it took the best part of the day to travel the autobahn. We arrived in Heidelberg late afternoon and I made my way to the Max Planck Institute, where Ulrich Häberlen offered to take us over to an apartment that he had found for us. The apartment was in a small village 9 kilometres outside Heidelberg, in a place called Schwetzingen. We were taken to a small apartment on Antoni Straße in Schwetzingen. This was on the second floor in a small block of apartments, about three apartments high, so that in total there were six apartments off a common entrance and stairwell. Ulrich had the key to the apartment and helped us take our belongings from the car up to the apartment, then bade us farewell.

My immediate thought when we got there was that we needed to get some shopping locally and something to eat, as we had not eaten since lunch. We simply locked up the apartment, got back into the car, and drove into the shopping area, but to our dismay we found that all shops were closed. It was a Wednesday, which was early closing, and absolutely nowhere was open. The only places that were open were little *Stube* or bar rooms, where you could have a drink of beer or wine and a light meal. There was such a bar on Antoni Straße, so we went back to this bar and had something to eat. Because the rules were quite different in Germany, it was possible to take the children into the bar. This was a fairly common practice at that time.

Alt-Heidelberg

Our eldest daughter Sarah was 5½ years old and had already had six months schooling in England, but in Germany schooling for children started at 6 years old. Nevertheless I made enquiries at the local *Grundschule* and found that it was possible for her to be accepted in the first year of the *Grundschule*, but Gillian at 3 years was too young and had to go into a kindergarten elsewhere. We found a kindergarten at a Roman Catholic school run by nuns. We took her there, but she was a little bit too young and did not like the look of the nuns, whom she thought were witches. Nevertheless we did try her once or twice at the kindergarten, but she really did not like it from the word go. Sarah, on the other hand, seemed to fit in extremely well in the *Grundschule*, and there were other children of *Ausländer* at the school, particularly Turkish children. Her teacher at the school was a Frau Natie. She could speak some English and she accepted Sarah in the class. Within weeks Sarah was speaking a little German and joining in with the games of the children in our block of apartments but also in school activities.

Eventually I turned up on my first day at the Max Planck Institute and found Professor Hauser and his secretary, Frau Türm, who introduced me to the various members of the group. I had already met Ulrich Häberlen, but I met others, including Hermann Brunner, Herr Stehlik, Herr Spieß, Herr Schweitzer, and the technician Herr Zimmerman. Of course there were others in the group whose names I cannot remember.

Shortly after I had arrived I started to interact with Peter Grannell in Nottingham, and we had a very intense and prolific correspondence. Because we were corresponding about new ideas concerning NMR diffraction, all my letters were handwritten, and I did not ask Frau Türm to type any of them, although her English was at such a high standard that she could easily have done so. One of the major problems that I noticed when arriving in Heidelberg was that virtually everyone in the lab could speak English rather well. I found this helpful to start with, but later on it became frustrating, because one of the reasons that I went to Germany was to improve my German, and it was extremely difficult to do so. The one person whose command of English was not as good was Hermann Brunner, and I was able to converse with him in German. He was extremely helpful as far as my language was concerned. He did speak English, but he seemed quite often reluctant to do so and was more willing to speak in his own language.

Häberlen's main interest at that time was in applying line narrowing techniques to a whole range of materials looking at the chemical shift anisotropies of protons in the various crystals with which he was working. I had mentioned to him some of the tricks that we were using back in Nottingham to stabilize the pulse sequences. One of them was a 90° pulse correction system; another was a phase compensation system, and this interested Häberlen because he was having similar instability problems with his rig. He suggested as a specific project that I build some equipment that he would incorporate into his apparatus in order to provide phase compensation and 90° pulse length compensation, thereby stabilizing his apparatus. While this was something that I had already achieved back in Nottingham, I discussed the details with Ulrich Häberlen, and he produced a circuit diagram that looked as though it would work well and fit in with the electronics of his system. I therefore pulled together all the necessary components, had a printed circuit made, and wired up what was essentially Häberlen's circuit. I tested it and it worked quite well and was therefore incorporated into his NMR apparatus. He also tried some of the pulse sequences that we had developed in Nottingham and found, together with the stabilizing circuit, that he was able to get marginally better line narrowing in the samples he was looking at. These results therefore led to a publication in English in the German Journal *Die Zeit*.

Meanwhile, back in Nottingham, Peter Grannell was making steady progress on the NMR diffraction experiments. By about March 1973 we were ready to write some of this multiple pulse diffraction work up as a paper for presentation at the First Specialized Colloque Ampère meeting, which was to be held in September of that year in Krakow, Poland. I wrote up the paper while in Heidelberg and sent it off for consideration by the local Ampère committee in Krakow. The paper was accepted and immediately elevated to an invited paper for presentation at the meeting. As it turned out, the paper followed the opening invited paper at this Colloque Ampère meeting, given by Raymond Andrew.

In Heidelberg I felt that I was not making sufficient progress in the lab with my German language. I decided therefore to take matters into my own hands and I announced to the group that I would be giving a talk on other work that I had been involved in back at Nottingham on NMR double resonance effects, but that this talk would be in German. This came as a big surprise to everyone in the lab, and when the day came for me to give the talk everyone sat dutifully and listened patiently to my broken German. After I had finished I fielded a few questions, which I was able to answer in German, and it was clear from the response of both the staff and the various research students that from that day on I wanted them to talk to me in German and not in English, and this they did dutifully. Shortly after this event I was invited by Dr Michael Mehring to visit his lab and give a talk on the double resonance work that I had been talking about in Heidelberg. So, with improvements in my presentation, I travelled to Bochum, and there gave my talk again in slightly improved German, and this was graciously received. These two talks were a turning point in my learning and practice of German, and from then on I noticed a very considerable difference in the attitude of people in the laboratory. Everyone wanted to speak to me in German with a few exceptions. Among the exceptions was Professor Karl Hauser himself, although eventually I did get him to speak to me in German.

An odd thing occurred shortly after we had moved into the apartment in Schwetzingen when we decided that, in order to meet people in the lab, it would be a good idea not only for me but also for Jean for us to hold a party in the apartment and invite all the people in the lab to come along. This way she could meet them and also make a social occasion of it. As we organized the event, we had ready acceptances from virtually everyone, but with the one exception of Professor Karl Hauser. He said he was unable to come to this party because we had not been to his apartment and he could not possibility come if we had not been to his house first. So he did not turn up, but shortly afterwards we were invited to his house and we met his new wife. We had dinner there, and it was a very enjoyable evening.

Back at home Sarah was making great strides with her German at school and in the afternoons when there was no school she was playing in the back yard of the apartment block with other children who lived in the apartment. It really was difficult to notice the difference between her talking and shouting in German and the German children there themselves, so she very quickly acclimatized and got into the language. Gillian, on the other hand, did not persist with learning German; she had thought it was very strange that nobody in her kindergarten could speak English, and as far as I could tell she never really understood that in Germany the children speak German not English. Jean had made some progress with learning German and she was eventually able to go out to the shops and buy necessary groceries and generally make her way. So her language was slowly improving. We had befriended people in the apartment above us, Ingrid and

Alt-Heidelberg

Thilo Bauer. But both Ingrid and Thilo could speak very good English and insisted on doing so on all occasions we were with them, so there was very little possibility for Jean to improve her German then.

At the lab, apart from Karl Hauser, the eldest person in the group seemed to be Hermann Brunner. His attitude to me, especially after I had given my talk in German, was to give me every encouragement by speaking to me virtually on all occasions in German, and this I found extremely helpful and useful. On the social side, he and his wife, Frau Brunner, were kind enough to invite us to their home on several occasions. On one occasion we went for afternoon tea at one of the many Chateau Restaurants in the region. The Brunners lived in Schriesheim, which is just outside Heidelberg. One afternoon Hermann took us up to the *Tingstadt*, so called, which was one of the arenas built by the Nazis and where Hitler had given speeches in the 1930s and during the war. When we got there, we found that the place was very much as it had been left. There were still large swastikas that had not been removed, as they had been in other places and on other buildings. When we visited the Brunners' house I noticed on the bookshelf a copy of *Mein Kampf*. It is a book that I tried myself to obtain while I was in Germany, but it was difficult if not impossible to obtain it at that time. Hermann's father had been a *Pfarrer*, a priest, I believe in the *Evangalische Kirche*, and was not sympathetic to the Nazi regime. Hermann told me that he had been a member of the *Hitler Jugend* only because it had been compulsory to join it and he really had no interest in it whatsoever, especially since, during the Hitler regime, the *Nazi Zeit*, children had been encouraged to report back on conversations overheard in the home.

While we were in Heidelberg, Jean and I visited the area around Schriesheim with the children. One weekend we came across a small uniformed Salvation Army group, some members playing in the band and others singing hymns. I noticed on their cap bands that they were members of the *Heilsarmee*. I approached one of the ladies and explained that we were from Nottingham in England, but that originally I was from Camberwell in London, not far from the William Booth Memorial Centre on Dog Kennel Hill, roughly half a mile from Camberwell Green. The encounter reminded me of my youth, when I had been a member of the Salvation Army Sunday School at Camberwell Green. There I had regularly attended Sunday School and in particular the Bible reading and study group. Subsequently, on our return to Nottingham, I learnt that William Booth originated from Nottingham.

The next most senior assistant in the Heidelberg group was Herr Dieter Stehlik, an amateur guitarist. I can still hear him playing and singing the famous German song *'Ich hab mein Hertz in Heidelberg verloren'*. Towards the end of my stay in Heidelberg, Dieter went off to take up a Chair in the Freie Universität in Berlin. However, I heard recently that he died suddenly in 2007. Another colleague, Hans Spieß, continued working in Heidelberg but subsequently, when he finished his Ph.D. studies, went off to take up a position and eventually a Chair in Mainz, another university on the Rhine. A third colleague, Dr Dieter Schweitzer, continued his studies but eventually left to take up a position as a professor down on the Bodensee in the University of Konstanz. He and his wife were living in Mannheim at the time and invited us over to visit them socially on at least one occasion, as did the Spieß family, as well as the Häberlens.

Travels further Afield

During our year-long stay in Germany we were able to visit Leipzig in the eastern part of Germany, and there we were invited by Professor Harry Pfeiffer of the University of

Alt-Heidelberg

Leipzig to visit for a few days and give a talk in the department. The trip to Leipzig required visas to get through the borders into and out of Leipzig. The sharp contrast driving in Eastern Europe as opposed to Western Germany was striking. There were virtually no cars on the road in the small towns and villages and you could drive for miles on the autobahns in East Germany rarely seeing another vehicle. Despite the severe difficulties and conditions that people were living under in East Germany at the time, Professor Harry Pfeiffer had arranged for us to stay with a family for a couple of nights while we visited the laboratory.

On another occasion we made a much longer visit, which took us through Czechoslovakia to Poland and eventually to Krakow. This was at the invitation of Professors Hennel and Blicharski at the Jadrowej Institute of Physics, ostensibly to attend a business meeting held on 11–12 June 1973 in connection with the forthcoming First Specialized Colloque Ampère Conference, which was then in the planning stage. On returning from Krakow we came back through the Tatra Mountains and into Hungary to visit Professor Kalmàn Tompa, at the Central Research Institute for Physics in Budapest. In those days it was necessary to have not only passports but also entry visas into these communist-controlled countries and we had been told that when we arrived at the checkpoint into Hungary the necessary visa would be available at the checkpoint. But when we arrived there was no visa waiting, so we were held up for quite a while with our two children while the border guards made enquiries. Eventually, after about an hour or so, we got permission to pass through.

One evening in Budapest Kalmàn took us to a famous restaurant where gypsy violinists serenaded us while we ate. It was all very romantic. On our way through Czechoslovakia to Prague we passed over a gorge. I glanced over the side and there down below was a river running and a notice on the bridge saying 'The Vltava'. It reminded me instantly of the Vltava Suite by Smetena, and when finally we got to Prague I found a music shop and asked about the works of the famous Czech composer. They had in stock copies of the suite *Ma Vlast*, my country, including the famous piece 'The Vltava' or in German 'Die Moldau'. I bought a compact disc recording that I still have today.

I found these trips scientifically extremely interesting. We visited people who were at the top of their profession in science and particularly in NMR, but in most cases they had no real equipment, only rather poorly made or poorly manufactured equipment, which meant that they were unable to perform quite a few of the experiments that we were lucky enough to be able to carry out both in Western Germany and also in Britain. The whole experience made me appreciate how fortunate we all were to live in the West and not under communism, where life was made extremely difficult for everyone because of constant shortages in virtually all the necessary supplies for living and certainly for virtually all the supplies required to run a laboratory and perform new and worthwhile experiments.

During the Easter break our mothers visited us in Heidelberg. I drove over to Frankfurt airport to meet them and took them to Heidelberg. My mother commented that she found it most strange—nobody spoke English. We had a most enjoyable time showing them the sights in and around Heidelberg. They stayed for a week, and the time seemed to fly by.

Other family members visited us, including my brother Sid and his first wife Margaret and their two children, Linda and Steven. They came by car and also stayed for about a week. I recall taking them to a famous pub in central Heidelberg, Zum Roten Ochsen. We all sat round a large *Stammtisch* drinking. When we got up to leave, a group, probably of students, stood up and gave us the Nazi salute and shouted 'Sieg Heil' as we left. I am sure it was a bit of fun on their part, but it was slightly worrying for Sid and his family. Other visitors included Laurie Challis and his wife, a colleague from the Physics Department at Nottingham.

Back in Heidelberg we had an unexpected visit from Professor Hennell, who, I recall, had visited the Bruker Company in Karlsruhe, and he was there to negotiate the purchase of a Bruker NMR system for his laboratory back in Krakow. That was a very pleasant and unexpected surprise for me. Later on we had another visitor, Professor Jiri Jonas. Although he and his wife were originally from Czechoslovakia, they had settled in Urbana, Illinois, and had come to visit us for a short stay of about three months or so. I had met him earlier when we were living in Urbana. Jiri was working in the Noyes lab in the Department of Chemistry and worked in Herb Gutowsky's group doing NMR, along with Bill Derbyshire, who was also there at the time.

As summer approached, the *Zeyer Grundschule* in Schwetzingen organized end-of-year examinations and Sarah's class was tested and she passed for acceptance for the following year. There were one or two other children in the class who were not so successful and were referred to repeat the class the following year. When we explained to Sarah's teacher, Frau Natie, that the end of my sabbatical period was approaching, she kindly obtained permission from the headmistress of the school to keep all of Sarah's schoolbooks. We took them with us back to England as Sarah was keen to continue with the language.

On Friday, 17 August 1973, we packed up all our belongings in the car, together with the additional items that we had acquired, which went on to a roof rack, and we left Schwetzingen, retracing our journey of a year earlier back on the autobahn to Aachen, where it was necessary to show our passports to pass into Belgium. I got out of the car to take our documents to the passport check point, and when I returned I got into the car and started to pull away. I suppose I got twenty or thirty yards away when I looked in the rear mirror and saw that our younger daughter, Gillian, had got out of the car when it was stationary and was running after the car crying out and we had very nearly left her behind. While we all had a good laugh about this afterwards, it was a great shock for Gillian, who at that time was still only 3 years old, and this episode remained imprinted on her memory for many years after.

Once in Belgium, our next task was to make our way through into France and eventually to Boulogne in order to pick up the ferry to return. While travelling in France we stopped for a bite to eat at a restaurant and to my amazement I found that when I tried to talk in French I couldn't speak a word. Whatever I wanted to say, German came out instead of French, and this I found to be a very strange experience indeed.

Our return Channel crossing was made via the hovercraft and took just thirty minutes. We were able to clear customs by around six o'clock that day and immediately got on the road and arrived back in Nottingham around four hours later that evening. To this day, thirty-nine years later, we still have the calendar, a plywood cutout of the *Schornsteinfeger,* the chimney sweep, with the tear-off date block attached below, showing the day that we left Schwetzingen—Freitag, 17 August 1973 (see Figure 23).

Back in Nottingham I had five working days and some frantic hours to prepare my talk, have the necessary slides made in the department's photographic unit, and make travel arrangements for the Krakow meeting. We decided to travel by coach. Much of the preparation of my talk had been done in Heidelberg, and it was only a few slides that needed to be made in Nottingham. These were assembled. We repacked our cases and, together with the children, we drove to London

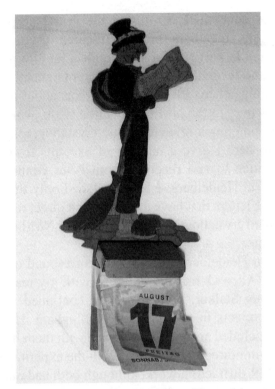

Figure 23 Departure day, 1973. The chimney sweep (*Schornsteinfeger*) calendar.

on the Friday preceding the conference. Our children were left in London with Jean's mother, while we travelled over the weekend, arriving in Poland in time for the start of the meeting.

Wechselwirkung zwischen Nottingham und Heidelberg: Correspondence with Dr Peter K. Grannell 1972–1973

I left Nottingham for a sabbatical year in Heidelberg at a critical time as far as the new idea of NMR diffraction in solids was concerned. So during the year I corresponded with various members of my research group back at Nottingham. In particular, I spent quite a lot of time writing letters to and receiving letters from Peter Grannell, who was just finishing off his Ph.D. work and was about to transfer his efforts over to the NMR diffraction work, which, in my letters, I sometimes refer to as the crystallography experiments. So here I give a number of excerpts from various letters written to and received from Peter Grannell while I was still in Heidelberg. I have included only the relevant parts of the letters that have a bearing or a direct reference to the so-called crystallography experiments or NMR diffraction experiments.

While in Heidelberg I did in fact correspond quite regularly with Allen Garroway, and also with my research student Dennis Stalker, both of whom continued with their research projects in my group at Nottingham. However, I have not included any correspondence with them here, since they were not directly concerned with the experimental side of the NMR diffraction work, although both had made some contributions and were therefore included in the paper that I presented at the Ampère conference in Krakow. Allen, of course, was to make substantial contributions to slice selection subsequently.

6 November 1972—University of Nottingham, Department of Physics.

P.K.G. As regards the crystallography experiment. J. E. Tanner, *R.S.I.* 36, 1086 (19), gives a geometry for coils to produce a homogeneous field gradient, in the presence of pole caps. See sheet (2b). Al pointed out to me that Murday and Cotts, *J Chem Phys* 48, 4938 (1968), obtained field gradients homogeneous to within 1% over 1 cm of sample in a 2″ magnet gap using a coil based on

Tanner's calculation. We should be able to manage that or slightly better.

13 November 1972—Max Planck Institut für Medizinische Forschung, Heidelberg

P.M. I was pleased to hear that some progress is being made with the crystallography experiment. Of course 1% accuracy in the field gradient is not very much, but I suppose it will enable the initial simulation experiments to be performed.

29 November 1972—University of Nottingham, Department of Physics

P.K.G. I shouldn't like to mislead you into thinking that I have done much at all on the crystallography experiment. I am about to put a bomb under the electronics workshop to get them moving on my receiver and new oscillator-gate-phase shifter. In the near future I hope to get the machine shop to make me a perspex coil former for the field gradient coils. But it is going to be after Christmas before I descend on the lab with a vengeance. This computing has taken much longer than I had hoped.

21 March 1973—University of Nottingham, Department of Physics

P.K.G. Progress on the field gradient experiment: I have hitched up Alwyn Finney's probe to my rig. The coils tuned well enough, but I reduced the number of turns per side on the transmitter coil from 4 to 2 to get a little more H_1. I tried both series and shunt tuning (using a $\lambda/4$ line for matching in the series case) but the shunt tuning gave best H_1. The transmitter coil with no added shunt damping resistor gives $\sim H_1 = 3.5$ Gauss, which is rather poor. The volume enclosed by the coil is rather larger than

I had in the double resonance probe and this accounts probably for the low H_1 and the low 1.3% inhomogeneity (water-rotating frame $T_2^* \sim 170$ µs). The homogeneity of H_1 may have been better but for the positioning of the connection from downlead to coil—this means that one side has about half a turn more than the other.

30 March 1973—Max Planck Institut für Medizinische Forschung, Heidelberg

P.M. I agree that the oddities in the rectangular sample decay arise due to deviations from the linear grad. The fact that a circular section sample takes up less volume would I think improve the experimental result—as observed. On the lamellar sample, I think you should have in the container volume about $1/3$ sample.

If the plate thickness Δ is thin enough, you could try stacking sample plates in pairs, trios etc., if the finest separation is not resolvable. In all cases, the mark/space ratio of ½ would be maintained by redistributing the plates.

By the way, I am thinking of mentioning these experiments in my talk at Krakow. I hope we have some tangible results soon!

Back on the field gradient experiments: If you get something working reasonably well, I think it would be a good idea to try a similar experiment on a solid, e.g. a multi-layer sample of CaF_2—perhaps you could discuss the possibility with Al. I think the solid would make a more convincing test, once you have the field gradient coils sorted out.

12 April 1973—Max Planck Institut für Medizinische Forschung, Heidelberg

P.M. I am looking forward to seeing the multi-layer sample response you mentioned in your previous letter. I have been thinking about a practical way of fabricating a multi-layer sample in a suitable solid and I believe I have a method which should work. The solid is adamantane, which we have looked at here with multi-pulses. As you know, this is a waxy sort of material that is not easy to pulverize. I have tried dissolving in CS_2 and it appears to be extremely soluable. The basic idea is to have made out of nylon, two multi-slotted spacers (if you look through my sample box, you will see a similar thing made from copper). The spacers (one each end) support a set of plastic wafers with uniformly spaced air gaps. Adamantane is deposited between the plastic plates by evaporating off the CS_2. This process is repeated until the spaces between the plates are filled.

I have calculated that for a $3/8''$ sample diameter with 10 plates and with mark/space of 1 and a field gradient of not greater then 2 G/cm the experiment should work on the multi-pulse rig. I don't think it matters if the support is made of protons, only the adamantane portions will give the line narrowing required. Do you think you could explain this to Denis and get him to have the thing made up *quickly* in the workshop? As I said in my last letter, I would like to speak about these experiments in Krakow and I am most anxious to have a result for a solid as well as a 'liquid'!

13 April 1973—Max Planck Institut für Medizinische Forschung, Heidelberg

P.M. Many thanks for your letter of 9 April, which I received today. Needless to say, I am delighted with the preliminary results you have sent me. They are, I think, better than you realized, because I think you were trying to interpret the time transients incorrectly. If you look at the theory for S_N you will see that for N slabs (NB n runs from $0 \to N$ so in your cases 4 slabs gives N = 3 and 3 slabs gives N = 2), if $2\pi/\gamma Ga$ then I would expect to see results as shown here [see Figure 24]. Both cases are

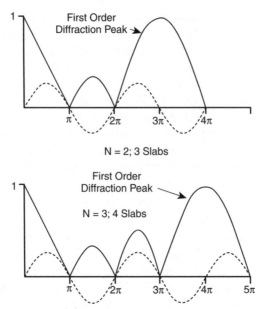

Figure 24 Normalized diffraction peaks for 3 and 4 thin slabs of material.

drawn for modulus of signal. This is exactly what you see in your photos!

On page 71 of your lab notes you have mistakenly compared the theoretical time ($t = 2\pi/\gamma Ga$) to the 1st order diffraction with the first peak on the photographs for the two samples. You should have compared the theoretical numbers with the 3rd peak for the rubber-paper and the 2nd for the rubber-paxoline. Using your measured times we then have:

rubber – paper		rubber – paxoline	
exptl	theor	exptl	theor
760μs	793μs	590μs	624μs

Considering the crudity of the samples, the agreement is excellent.

You will notice from the theory that the 0th and 1st order diffraction peaks should have a width ~2 x the intermediate peaks. This is observed experimentally in the photos. In my notes, I inadvertently indicated on the sketch the half-width of the 0th order diffraction peak. This is wrong. It should be the full width. This comes from setting $½\gamma Ga(N + 1)\Delta t = \pi/2$ which gives $\Delta t = \pi/\gamma Ga(NH)$. We further notice that if t_1 is the time to the 1st order diffraction max then $t_1/\Delta t = 2(N + 1)$.

For the rubber–paper sample $N = 3$, $t_1/\Delta t = 8$.
Experimental value is unclear.
For the rubber–paxoline sample $N = 2$, $t_1/\Delta t = 6$.
The experimental value 600 µs/120 µs = 5.

So I should say here again, the agreement is excellent. In order to see patterns similar to my original sketches i.e. the subsidiary peaks small c.f. 1st order diffraction peak, one must use large N. For a perfect uniform gradient, one could go on indefinitely increasing N. Your FT's already show that the off-axis gradients in the field coils are already broadening the spectrum.

My guess is that with the present gradient coil, you could probably go to 8 slabs. Expect a just resolved structure. Can the gradient coils be improved?

7

Krakow and the Lauterbur Epiphany

Revelations in Nature

The first Specialized Colloque Ampère was held in Krakow from 28 August to 1 September 1973. Jean and I travelled by coach, with a stopover in Leipzig. The conference opened with Raymond Andrew's paper entitled 'High Resolution NMR in Solids', followed by my paper entitled 'Multi-Pulse Line Narrowing Experiments: NMR "Diffraction" in Solids?' I spent the first ten minutes or so talking about multiple pulse techniques and line narrowing in general. Then I concentrated on the new results that we had obtained on the sets of three and five camphor plates with and without a gradient, showing the diffraction effects in k-space and the resolution of the three plates with and without a gradient in real space (r-space). The results were well received. At the end of the talk there was some time for questions. One of the questions was raised by none other than my old adversary, John Waugh. He asked whether I was aware of what he thought was similar work that had been published by Paul Lauterbur a month or so earlier in the journal *Nature*. I said that I was unaware of any similar work and had not seen the article in *Nature*. In fact it was not a journal that I read regularly. I asked whether the work had been done in solids using multiple pulses. John's answer was that it had not been done that way. The work was concerned with producing an image from test tubes filled with liquid. Naturally I was intrigued to hear about the work of Paul Lauterbur and could not wait to get back to Nottingham to visit the library and look for myself at the paper that had been published. When we finally got home after the conference I checked the paper in the library and was struck by the contrast between Lauterbur's approach to imaging and our approach. Paul Lauterbur had used a c.w. imaging technique, which worked in real space by rotating the sample and taking a number of projections of the object at a series of angles relative to the fixed gradient direction, whereas we had used the k-space approach and introduced all the necessary mathematics in k-space, which bore a very strong similarity to standard diffraction theory as I had alluded to in my talk and in the title. To the casual observer the two approaches looked to be unrelated, and in retrospect it was most astute of John Waugh to have noticed the similarity between the two imaging approaches, especially, it has to be said, since the results of the two approaches were on different materials—namely, solids, on the one hand and liquids, on the other. Apart from the conference proceedings that were published as part of the Specialized Colloque Ampère Conference, we also sent off a paper to the *Journal of Physics*. A letter in that journal was published shortly after, around October 1973.

It was obvious to me, even at that early stage, that imaging in solids *per se* was significantly more difficult than imaging in systems where the NMR line width was already very narrow—namely, in liquids. The multiple pulse technique that produced the line narrowing in a solid material was an additional complication that we really did not need. It was decided fairly early on that maybe that was something that was perhaps ahead of its time and could wait until later to be further developed. In the meantime we should perhaps concentrate on liquid systems. Although my paper had created a great deal of interest and discussion in Poland, when we returned to Nottingham there was virtually no mention of it at all, although there may well have been people from Nottingham who heard me speak in Poland.

A Sensitive Point

Early in 1974 a group of NMR people from the Physics Department at Nottingham went off to India to an International Society for Magnetic Resonance (ISMAR) conference in Bombay. Among the group travelling were Dr Bill Moore, a Lecturer in Physics working in his own electron spin resonance (ESR) group, and Dr Waldo Hinshaw, an American postdoc, and Dr Bill Derbyshire, who were both working in Raymond Andrew's NMR Group. At the conference they heard Paul Lauterbur speak about his work on imaging, and apparently on the aeroplane back Bill Moore and Waldo had a long discussion about Paul Lauterbur's work and discussed what they thought could be an improvement to the technique. Between them they came up with the idea of sensitive point imaging.

In the weeks following their return to Nottingham, they were rather secretive about their new proposals, until it transpired that they had filed a provisional patent with the National Research and Development Corporation (NRDC). Once that had been done, Waldo Hinshaw set to work to build a system that would allow him to produce an image by the sensitive point method. This proceeded fairly rapidly, and within six months or so he was able to produce cross-sectional images of various small objects placed in the magnet, including a chicken leg and various plant materials.

Slice Selection

Since the Krakow meeting I had been increasingly concerned about the problem of selecting an initial slice. This was also a problem with Paul Lauterbur's method. He had defined an imaging sequence but had not defined a way of selecting a slice. The whole process of slice selection was one that had worried me for some time, and in late 1973 I started to think very seriously about this problem. Peter Grannell and I had many discussions about this. Allen Garroway had also shown some interest in imaging, and he started to think about slice selection. Peter Grannell and I had proposed two methods of slice selection. One method used a selective saturation sequence in which everything but the line of material of interest was saturated. The sequence was followed by a 90° pulse that allowed observation of the unsaturated region of the spin system. In a second method, we selectively excited the active slice of interest and followed that by looking at the free induction decay following the selection pulse. A similar technique was developed independently by Allen Garroway, but in his case he used a sequence of short pulses to achieve selective excitation, thereby effectively defining a slice of material. The three approaches looked to be roughly similar with similar effects. We all thought at the time that perhaps the sensible approach would be to combine all ideas and methods into a single publication and publish this. But before doing so we thought it would be sensible to take out a patent to cover the general question of slice selection and combine that with an imaging technique based on line scanning techniques. So this is what we did.

The patent took a month or so to prepare, and it was necessary to perform some experiments to establish the validity of our approach or proof of concept. These were carried out, and the results of this work finally led later to some publications. Both the experimental work and the patent material were written up and submitted to the NRDC. This was filed shortly after Waldo Hinshaw and Bill Moore had filed their patent.

Our patent details included slice selection in arbitrarily oriented planes and were an attempt to broaden the claims in the patent. The paper was published in the *Journal of Physics E: Scientific Instruments*, 9 (1976), 271–8, and included some cross-sectional images of plants that I had picked from our garden in Chilwell.

There followed a flurry of papers that we produced using line scan imaging. Later on I thought about human imaging and what we could do to achieve this. The coil system that we had been using was really part of the original NMR apparatus that we had built—namely, the computer-controlled equipment (see Figure 25). The operating frequency was 9 MHz and the sample size was 1.5 cm in diameter, and therefore the only living object that I could think of on the spur of the moment was a human finger. Could we reach in and place a finger inside the coil while we were imaging? It turned out that my fingers and those of most of my students were too large and it was impossible to get the index finger

Figure 25 Terry Bains, Peter Mansfield, and Andrew Maudsley, c.1974. Terry is leaning over the Honeywell computer.

of a large hand comfortably inside the coil. The exception was one of my students, Andrew Maudsley. He had particularly thin fingers and could get his hand and therefore several fingers, including his index finger, separately inside the coil, allowing several of his fingers to be imaged comfortably.

Line Scanning

In line scan imaging a whole line of data is obtained in a one-shot experiment, which means that, for example, a 64^2 data array could be obtained in 64 times the imaging time for a single line containing 64 points. This is to be compared with a point scan imaging time of 64^2 times the time to obtain a single point. Thus it was clear to me that the idea of point-by-point imaging was always going to be much slower than line-scan imaging, and indeed as time proceeded it became apparent that by modifying the imaging technique one could obtain not just one line at a time but a plane at a time, and these arguments had clear implications for medical imaging. My own ideas on the topic of medical imaging were always concerned with the speed of imaging, and right from the word go this was something that I emphasized in any talks that I gave or indeed in any papers that I published. This constant emphasis on imaging time began to upset the other group in imaging at Nottingham, which, shortly after the Indian conference and the filing of the patent, became a subset of Professor Raymond Andrew's NMR activities.

8

Hounsfield and EMI

Central Research Laboratories

One of the flurry of papers authored by me and Peter Grannell was picked up by people in the Research Division of EMI in London. When the paper was published I received a letter from a Dr Hugh Clow, who invited me down to the EMI Central Research Laboratories in Hayes, Middlesex, so that he could find out more about our imaging methods. I drove down to EMI and was met by Alan Blay, who was in charge of a small research group that included Hugh Clow. Their main research interests were ostensibly in the manufacture of magnetic recording tapes, but they were quite interested in magnetic resonance imaging because EMI had recently entered into the medical imaging business through the work of Godfrey Hounsfield. He had recently invented an x-ray technique based on computerized tomography and had published work, the details of which I later discovered. I was naturally very keen to meet him.

When I arrived at EMI, I was ushered into a small seminar room. I gave a talk on MRI and presented the results that we were producing at the time. I also spoke, to some extent, about our future plans. There must have been a dozen or so people in the room, and I was not introduced to all members on that occasion, but I do remember meeting a Dr Bill Percy, who was introduced to me as the theoretician of the group. My talk lasted roughly an hour, and Alan Blay had questions that he asked about our proposals and the details of our methods and so on. The whole business was over, I suppose, by

4.30 or so in the afternoon. I asked whether I could meet Godfrey Hounsfield, but was told that he was not available. Alan explained that the EMI management thought it would not be a good idea for me to meet Hounsfield, since hearing about a new imaging modality might distract him from his work on computerized tomography. I collected my slides together and prepared to leave, accompanied by Alan Blay, but as we walked along the narrow corridor to the main entrance to the Central Research Laboratory a man came towards us. I stopped to make room to let him by, and at that point Alan Blay turned round and said: 'Oh, may I introduce you to Godfrey Hounsfield?' There he was standing in front of me in the corridor. He was very surprised and said: 'Who are you and why are you here?' and I said that I had just given a seminar on NMR imaging, and I was surprised that he had missed my talk. He was also surprised. In his response he said: 'I would have loved to have been there but nobody told me anything about your talk.' So we started chatting in the corridor. Alan Blay tried to usher me forward to leave, but Godfrey Hounsfield was quite interested in what I had to say. So at that point he asked: 'Would you like to come along to my office and we can have a chat about things and perhaps you could fill me in on exactly what NMR imaging is about?' So we went along to his office, which was just along the corridor. It was by that time close to five o'clock. Alan Blay made his farewell. We sat in Godfrey's office and in effect I gave my talk all over again but on a one-to-one basis, and Godfrey was absolutely fascinated with what I said. I suppose I spent about an hour and a half to two hours going through the details. It took that long because he stopped me every few minutes and he wanted to know why this and why that and so on. So it really ended up being a two-hour personalized tutorial. Every little idea that was mentioned with which he was unfamiliar Godfrey wanted to know about. That was how I met Godfrey Hounsfield (see Figure 26).

Figure 26 Godfrey Hounsfield with me. A poor double exposure, but this is the only photograph that I have of Godfrey Hounsfield.

I finally got away that day at about seven o'clock in the evening. By that time, of course, most people had gone home, and only the porters and security people on the front desk remained in the building.

In the weeks that followed I received a letter from EMI inviting me to be a consultant on NMR and NMR imaging, which I readily accepted, and as a result of this I was invited to visit the Central Research Laboratories on a fairly regular basis.

The second time I met Godfrey Hounsfield was when he was invited to Nottingham to present a talk on computerized tomography (CT) imaging in the Physics Department. Invitations to the talk were extended to all departments and to the Medical School. The talk was held in our largest lecture theatre and it was full to overflowing on the day. Professor L. Challis gave the introduction to the talk, but he said to me prior to the talk that he was not sure whether a talk on X-ray CT scanning was entirely appropriate for a Physics Colloquium.

I disagreed quite strongly with this sentiment. I recall meeting Godfrey prior to his talk. He arrived at around two o'clock and had not eaten. I offered to get him something to eat, but by then restaurants on campus were closed. I then suggested that we might find a café outside the campus. I drove him towards Chilwell. We found a local café and stopped there for a meal before returning to the Physics Department.

Godfrey gave his talk and acquitted himself very well and was much appreciated by me and the audience in general, which reassured me that Laurie Challis's comments had been somewhat misplaced.

Godfrey Hounsfield and Allan Cormack shared the Nobel Prize for Physiology or Medicine in 1979. Godfrey died in 2004 aged 85 years. He was from Newark, Nottinghamshire. Allan Cormack died in 1998 aged 74 years.

From the CT scanning work that was going on at EMI I knew that they were able to produce colour-coded images of x-ray scans and they had the software available to do this. So on one of my visits after we obtained our NMR imaging results on Andrew Maudsley's finger I asked whether they could take paper tapes of the data that we had and turn them into coloured images for presentation on slides, which they agreed to do. As a result of this interaction I had about half a dozen rather nice coloured slides which I was able to use at conferences. In addition I was able to print the data as coloured images.

The MRC Enters the Frame

In the early 1970s, the Medical Research Council (MRC) convened a meeting of interested MRI groups to discuss their strategy in the possible funding of the then extant MRI groups in Britain. Among those groups were those in Nottingham and Aberdeen. Aberdeen had had an interest in ESR but had now developed an interest in MRI. Before the meeting, held at the MRC offices in London, I had the images enlarged and printed on photographic paper. These were then pasted on to A4 size boards. I took these down for distribution at the meeting while I gave a talk on our results and our methods of imaging. The meeting was chaired by Sir Rex Richards. I was also able to project slides onto a screen to maximum effect. I believe my talk had a huge effect on the members of the meeting and was also a tremendous shock for Raymond Andrew and the rest of the Nottingham crowd as well as the people from Aberdeen, none of whom had the slightest idea that we had already produced images of live human fingers.

One of the slides that I showed was an image crosssection, which had been drawn and annotated by Professor Rex Coupland, who had seen our images and was himself greatly impressed. So far as I was concerned, the meeting was extremely successful, and I was invited to submit an application for an MRC grant. Subsequently an application was put together and shown to Professor Andrew as Head of the Department for his comments. Andrew sat on my grant application for about six weeks, and in the end I went along to demand to hear his comments and to have the application returned. In the meantime, unbeknown to me, Andrew had put together his own MRC grant application and submitted it himself to the Medical Research Council. He waited for the outcome of his own application before returning my document to me with any comments. This delay infuriated me, and was the start of internecine squabbling and contention between my group and Andrew's group. As it turned out, Andrew was awarded a grant, and it transpired that he had asked for money to build an intermediate size imaging system that could image objects up to approximately 10 cm in diameter, whereas I had asked for money to make the jump from finger size imaging up to whole body imaging.

My grant application eventually went forward but because of the delay in submission was not considered until the next round of grants. It was awarded finally the following year, in 1975.

I must have impressed Rex at the meeting convened in 1974 to discuss the new requirements of all those interested in the development of NMR imaging, because later on in 1976 Rex invited me to lunch at the Royal Society and said that he wished to support my case for Fellowship of the Royal Society. Naturally I felt greatly honoured at the proposal, but it took something like seven years (the maximum allowed) before I was elected to the Fellowship in 1982–3.

It was during the course of 1976 that Raymond Andrew convened a meeting in Nottingham of interested people in imaging, including representatives from Aberdeen University as well as people from EMI. In addition we had one visitor from Zurich, Professor Richard Ernst (see Figure 27). Most attendees brought us up to date with their images and gave us short talks on the goals that they were pursuing. Although my group had made considerable headway in a whole range of topics, I chose to speak about an entirely new imaging method that I had worked out theoretically but for which I had really no experimental results. The technique was called echo planar imaging (EPI), a condensation of planar imaging using spin echoes. I spoke for something like half an hour, talking about this in great detail, and at the end of the talk the audience seemed to be left in stunned silence. There were no questions, there was no discussion at all, and it was almost as though I had never spoken. In fact I had given a detailed talk about how one could produce very rapid images in a typically one shot process lasting, conservatively, for something like 40 or 50 milliseconds.

A. Jim Hutchison   B. Andrej Jasinski   C. Paul Lauterbur   D. Raymond Andrew
E. Andrew Maudsley   F. Neil Holland   G. Rob Locher   H. Bill Moore
I. Richard Ernst   J. Waldo Hinshaw   K. Ian Pykett   L. Peter Allen
M. Hugh Clow   N. Bill Derbyshire   O. Peter Mansfield   P. Bill Vennart
Q. Joris Creyghton   R. Terry Baines   S. Peter Morris

Figure 27 The first MRI meeting held in Nottingham, 1976.

It was at this meeting that we learnt about the technique proposed by Kumar, Welti, and Ernst, and presented by Richard Ernst. This was a two-dimensional multi-shot technique and, although much slower than EPI, nevertheless seemed to have a great deal of potential for high-quality imaging for the future.

In the months following this meeting, Andrew's group acquired its laboratory magnet and continued to work on its 10 cm size imaging, and in the course of time produced some images of a live rabbit's head. This work was presented at a meeting of the British Institute of Radiology and, by all accounts, at the end of the talk Andrew received a standing ovation. This response apparently upset one or two people in his group, who felt that it was they who had done all the hard work. The group was not happy at the focus of attention created by the standing ovation.

The MRC Signals Support

Eventually I received approval from the MRC to go ahead with my plan to build a whole body MRI system, but it took some while to get properly started on this project. In the first place it was necessary to talk in much more detail to the Oxford Instruments people about the design and specification of the magnet. In addition it took some considerable while for them actually to wind each coil and make the complete magnet system. This went on during the early part of 1977, and the magnet was constructed during the latter part of 1977.

The magnet arrived eventually on the last working day before Christmas 1977. Actually it turned up in the late afternoon during a Christmas party at which members of staff were celebrating. I was called away to receive the magnet, which was loaded on this broken-down old truck with the back springs virtually shot. The truck was eventually reversed into our loading bay and the crane that was installed in the loading bay was used to take the coils off one at a time, together with the various bits and pieces that went into the magnet support assembly. This was done with me and one or two of my students whom I had managed to drag away from the Christmas party. Eventually the whole magnet was unloaded. However, it being the last day before the university closed for the Christmas break, there was very little we could do in terms of assembly. All that had to wait until the New Year had arrived and the department reopened again.

So it was the year 1978 that saw tremendous activity on our part to get the magnet parts assembled in one of the laboratories, which we had cleared out to receive the magnet and which was to be the designated spot where experiments would be done. All this created a flurry of activity in the New Year and the weeks that followed.

Blacksburg

The real pressure was on to get something up and running, so that we had enough time to do some preliminary experiments in order to present a paper at the upcoming Experimental NMR Conference (ENC) meeting in the States, which was going to be held in Blacksburg, Virginia, in April 1978.

We soon had the magnet put together, and called the specialists from Oxford Instruments to come up to Nottingham and align the coils to give the best magnetic field uniformity. This they did, and within a week or so the magnet was switched on and running quite well. The task then was to start to think about much larger receiver coils and to consider exactly how we were going to receive the NMR signal and how we were going to produce the RF power required to do the experiments. As it turned out, the RF power was really not a major problem, because in the adjoining laboratory we had a home-built multiple pulse rig, which had been designed to do experiments in solids. This had a very high power RF output of several kilowatts and could produce for protons one microsecond 90° pulses. In fact, the power we required for imaging was much lower than this. The real task was giving sufficient thought to the problems concerning the size of the RF transmitter coil and the tuning of the receiver coil.

This, of course, had all to be done at 4 MHz, but we had never before built coils so large. There was a period of experimentation with coil winding to see just what we could get away with in terms of coil size that would resonate at our operating frequency. Within a week or so we had produced crude signals that were subsequently improved and we were able to put large objects into the coil system and produce images of these objects. Generally speaking, we were examining, in many cases, cadaver tissue, which was placed in the large coil. In some cases we were using the whole storage jar in which preserved material came over from human morphology. In other cases the cadaver tissue was immersed in formalin to preserve it. In other cases we were looking at small unimmersed parts of the body. In all these cases the images we were able to obtain were of varying quality, and on occasions it was quite difficult to make out exactly what it was that the image represented. This was a major worry for me, because I began to get concerned that maybe the differentiation in signal strengths among the various tissues in the body would not be sufficient to give clear demarcation of the organ margins. On one occasion we had a torso brought over from human morphology and again the detail that was observable in that was really not clear enough to differentiate much in the way of the various organs and tissues. It was of some concern if living tissue were to behave in the same way, for then it would be a worry that we could have a problem with tissue differentiation.

We persevered with testing using cadaver tissue and also phantoms contained in glass tubes and containers, but time was pressing on, and I was worried that we would not have the key piece of information that we sought—namely, an image of a live human subject. In order to clean up the NMR signals we surrounded the magnet with aluminium sheeting to create an RF screen. This formed a small housing with its own entrance door. The day before we were due to leave for the Experimental NMR Conference (ENC), Peter Morris and Ian Pykett set up the machine, and I offered myself as our first whole-body human volunteer. I climbed into the machine and asked Peter and Ian to operate it when I called out that I was ready to go. The door was closed and of course it was pitch black inside. We hadn't had time to install a light inside the aluminium screening. The problem was that we did not have enough time to put a fully screened power cable inside. One could not just take an electric light bulb on the end of a lead and light the inside of the magnet housing, because the wire going in would have acted as an aerial taking extraneous signals from outside the aluminium screened enclosure right to the receiver coil. As a result, I was shut in the magnet enclosure in complete darkness. The electromagnet was a four-coil structure comprising two large coils and two smaller coils so spaced as to be a crude approximation to a spherical coil system. There were gaps between the separate coil sections. The arrangement that I thought at the time was easiest to try first of all was one where I stood between the two central coils so that the axis of my body was at right angles to the static magnetic field. I had around my waist a small elliptical coil, which had been tuned to 4 MHz. The problem was that standing in the magnet I really couldn't touch the coils, because after about ten minutes they got really quite hot. They reached about 55°C. My face was between the two large coils and rather close to the hot surface, so I had to be very careful that I did not burn myself on the coils.

I should say that about one week before we were ready to start this scan I had received a note from Professor Tom Budinger, who was a medical scientist working at the University of California in San Francisco. He had made a calculation that suggested that the gradient strength that we were planning to use on my abdominal image would be dangerous. He said in his note that this low field gradient strength could be enough to trigger cardiac fibrillation. I, on the other hand, had done my own calculation and did not really accept Tom's result. So, when I positioned myself in the magnet to be imaged, I called for the first pulse, just one pulse. There was a loud click from within the magnet, but I did not feel anything, so I signalled to Peter and Ian to carry on the scan. In fact I was in there for fifty minutes and outside my wife, Jean, and Peter's fiancée, Elizabeth, were standing by ready to pull me out if there were a problem. Hopefully they would have resuscitated me had there been a problem. In fact the scan went well and after fifty minutes and sweltering heat I got out of the machine dripping like a wet rag. We looked at the scans that were on the screen and decided that they were OK to present at the meeting, but we needed photographs to make slides. The results were quickly photographed, but there just was not time to have the film developed and the slides made. Peter Morris and I travelled to Blacksburg. I took the roll of film with me to America in my luggage and when we got to the States I found a small photographic shop quite close to the Conference Centre and asked the assistant if he could give me twenty-four-hour processing and get the slides mounted. This he did, and everything came together just in time for me to give my talk at the conference.

On the first night of the conference there was a social event, when invited speakers and organizers got together for a party. There was a fair amount of alcohol available that evening, and I did indeed have a few drinks. The next day I didn't feel too well and on the day after I was feeling even worse. My condition could have been exacerbated by the small quantity of drink that I had imbibed that first evening, but I began to worry that I was suffering from some after effect that had developed following my scan in the magnet. This feeling persisted for the whole time that I was at the conference. I gave my talk, which was reasonably well received, but the strange feeling that I had continued with me, and I began to wonder whether I should continue my tour, since following the conference I was scheduled to go on to visit another university centre in the States and give a paper there before returning to England. In the end I decided to try to carry on with my schedule and went to North Western University and met Athol Gibson, who had invited me to the department to give a talk. But by then I was feeling so ill that when I arrived at his house I had to go straight to bed. I was due to speak in the department the following day. When I got up the next day I decided there and then that I would have to leave and asked if Athol could make my excuses. So I returned to Nottingham a few days early. I still felt quite ill and went to bed when I got home. Eventually, a day or so afterwards, I went to see my GP and related the experience to him, but he was unable to help me. His attitude was that I was the expert on MRI and I should know if it was going to be dangerous or not. Of course, nobody knew the definite answer to that one, but it did not make me feel any better. In fact it seemed to prolong the state of anxiety that I had, which persisted for the next month or so. Eventually the anxiety feelings faded, but on the odd occasion they returned, especially if I heard something untoward connected with MRI.

In 1979, or thereabouts, I had a letter from John Waugh inviting me to write a review article for the series that he edited published by Academic Press. I accepted the invitation and started to pull together all the ideas that we had had on MRI, but because of my experience I also started to gather all the information that I could into a separate chapter on the biohazards of magnetic fields and the potential problems with MRI. In fact the article grew and grew, and it got to the point where I decided that, in order to finish it off, it would be better if I were to ask my research assistant, Peter Morris, who had by then submitted his thesis and was working with me as a postdoc, whether he would be prepared to work with me to finish off the book. So it was that the book *NMR Imaging in Biomedicine* was completed with Peter making a substantial contribution in the final version of the book. This was published by Academic Press in 1982.

In the meantime Raymond Andrew had been approached by the Wolfson Foundation, or rather the university had been approached to submit a grant application, and the Vice Chancellor of the University at the time, Professor Butterfield, a medico who later took the post of Regis Professor of Physic at the University of Cambridge, was quite keen for us to take up the challenge. Raymond wrote to me asking if I would like to be co-applicant on this grant. I readily agreed to this, and my name went on the application. But once the application had been submitted I heard no more about it. The Wolfson grant was made eventually, but unfortunately I was never invited to participate in the work, which was a shame, as it might have been an opportunity to bring the groups together.

9

The Golden Years

Head Images

Considerable progress was being made with MRI at EMI. For their first approach they decided to obtain a magnet similar to our electro-magnet and operating at 4 MHz for proton resonances. At that time no other group had decided to try imaging a human head, so the EMI group settled, in the first instance, for imaging Hugh Clow's head. On one occasion while I was there on one of my fairly regular visits I was told that the group had to be expanded to include some of the electronic engineering people in what had up to then been a separate support group. During this group meeting a person entered the room who turned out to be none other than Ian Young, whom I had met previously during my vacation job at E. R. Watts. He had changed jobs and at that time was working as one of the engineering support staff on the electronics side at EMI. What I did not realize until much later was that EMI had also approached Raymond Andrew and his group and engaged them too as consultants for the group at EMI. It was interesting, however, that the EMI group, in their decision to go for a head scan, decided to use the slice selection principle that we had conceived, and on the very first scan I believe they used a projection reconstruction technique, since this was something extremely familiar to the people at EMI, particularly on the CT scanning side.

Meanwhile, back at Nottingham, Raymond Andrew managed to get his Wolfson Foundation Grant and decided to buy a magnet. In his case he went to the same American company from which Paul Lauterbur had obtained his magnet—namely, Walker Scientific. Again the magnet operated at a field close to 0.1 T with an operating resonance frequency for protons of 4 MHz. In the course of the following year Andrew and his group were able to obtain human head images using that magnet system. But the technique that they used was the sensitive point approach, which was a relatively slow approach, as outlined previously. They were nevertheless able to obtain what were for the time quite good images. They were also able to look at a number of volunteers and a few patients. Once the machine was up and running there was never at any point a question of my being part of their imaging project. All discussion that we had had previously concerning collaboration was quietly and conveniently forgotten.

It was shortly after Bill Moore and Neil Holland had obtained their first medical image of a patient's head that disquiet in the group began to emerge. Apparently Raymond Andrew was giving a talk and wanted a particular slide of a head scan; so he went down one Saturday morning when the lab was deserted, had a look around, found the slide he wanted, and simply used it at a conference. Word got out later, and this caused some considerable concern for Bill Moore and Neil Holland, since it turned out that the image that was being shown was one of their best and most recent head scans. So, once again, Raymond had managed to scoop the best results from the group and present them at some conference or other, and the group was left very irritated about this. Because of Raymond's action, Bill and Neil decided between them that the work they were doing on the 4 MHz imager would no longer include Raymond Andrew. He was excluded from any of their further research work, and no experimental results were made available to him. This created something of a hiatus within the Andrew group, and as a result there were from that time three groups in the department, one headed by Bill Moore, one headed by Raymond Andrew and the small-scale 10 cm imaging system, and, of course, my group.

At around this time one of Raymond Andrew's students, Paul Bottomley, an Australian who had been working on the small-scale imaging system, had done some theoretical calculations concerning the radio frequency penetration that one could expect in tissue. His work, which I believe was presented in his Ph.D. thesis, showed that it would be increasingly difficult to produce images at frequencies much above 4 MHz. The reason he gave for this was that body tissue was sufficiently conductive to produce a penetration depth effect. However, the work that was being done by Bill Moore and indeed by ourselves showed that that was clearly not the case. If there were to be a limit, it would be at much higher frequencies. Peter Morris attempted to calculate what that limit might be in our 1982 book *NMR Imaging in Biomedicine*. A figure of 10 MHz was suggested, together with the assertion that the undoubted image corrections at higher frequencies would be a problem best avoided.

The split in Andrew's group came at a fortunate period for me, because it meant that the squabbling and difficulties that had gone on between Raymond and his former colleagues took the heat off me and my research operations for a period.

Electronic Support

My early work on multiple pulse NMR in solids was supported by an SRC grant, which included funds for an assistant experimental officer in the person of Terry Baines, an electronics engineer. Virtually all the new electronics and any computer control circuitry that was needed were designed and built by Terry Baines for our new pulse spectrometer. When imaging came along, Terry turned his considerable talents to assisting us on the imaging side for a while, until he left to take up a further position. I believe he went to the University of Cambridge to work in the Physics Department. Eventually I managed to get a replacement research experimental officer. His name was Ron Coxon, who was also an electronics engineer with considerable experience in a whole range of topics, including radio frequency work and especially pulse techniques and computer electronics. He stayed on in the group until 2008, during which time he went on to build several more systems.

Siemens and Erlangen

In the early days of the development of MRI, I was offered a consultancy by the German company Siemens and invited to their works and lab in Erlangen close to Nuremberg. In fact the offer had arisen through an old acquaintance, Alex Ganssen, whom I had met at MRI conferences.

The invitation came from a Dr Schittenhelm, who was in the process of organizing a small research group in Erlangen. When I arrived in Erlangen, I noticed that the lab was on Henke Straße (Hangman's Street). I was introduced to a number of German colleagues, and, of course, Alex was also present. They later took me into Nuremberg for lunch. The restaurant he chose was called *die Karze*—apparently a kind of detention house for recalcitrant students in bygone days.

Back in Erlangen I was asked if there were any promising students that I could recommend from Nottingham, as it was planned to expand the NMR group. As it happened, my very first student in MRI, Andrew Maudsley, was at the end of a one-year postdoctoral period working with Professor Ernst in Zurich. I suggested that he might be interested in joining the new group in Erlangen. He took up the offer in 1977–8, and the group got underway designing Siemen's first prototype MRI machine.

Later in the afternoon I was introduced to a number of people who would form the magnet design group. Among these was a Dr Schmidt, known to his colleagues as Rakete Schmidt. In conversation with him I later learned that he had worked at Peenemunde, on the design and construction of the V-2 rocket. He, of course, knew Von Braun very well. It further transpired that he had been a member of the pre-war society Die Gesellschaft für Raumschiffahrt (GfR), the Space Travel Society. And it was from this society that he knew Von Braun and the society's founder, Professor Herman Oberth.

He was intrigued to learn that I knew something about the pre-war activities of the society, and was especially interested in my period working at RPD Westcott. I think I had picked up my knowledge of the GfR from the book by Willie Ley entitled *Rockets*. The following day Rakete Schmidt brought in for me an original reprint of a paper of his published in the society's journal in around 1935.

While in Erlangen, Dr Schittenhelm mentioned to me that one of his hobbies was collecting different varieties of edible mushrooms. Sadly, I heard a year or so later that Dr Schittenhelm had died, of all things, from mushroom poisoning.

By pure coincidence, many years later, while viewing a rather grand house in Market Harborough, I spotted Willie Ley's book on the bookshelf in the library. The house owner, a 90-year-old brigadier, had died, and the house was up for sale. I offered to buy the book, but unfortunately it was considered by the Property Agents to be part of the house contents and could not be sold separately.

The Golden Eighties

As the 1980s progressed it became increasingly clear that the government of the day was interested in reducing the costs of universities by introducing and operating an early retirement scheme for many staff. This scheme was a voluntary one, and anyone who served on the university staff could volunteer to leave, but the conditions and the incentives to leave or take early retirement depended very much on one's age. If one were relatively young, and had just joined the university, then the financial inducements to leave were much higher than if you were a senior person with a relatively short time in post before retirement. Now Professor Andrew was in this second group. I remember one day his coming back to the department and saying that he had been to talk to the Vice Chancellor, Dr Weedon, about something or other. It emerged that the Vice Chancellor had told him in no uncertain terms that he wanted him to take early retirement. We were all surprised and bemused by the whole process, but as the early retirement scheme broke more widely, it became obvious that there were a number of people throughout the University of Nottingham who were willing to leave. In particular there were in the Physics Department a few people who took voluntary redundancy and were suitably compensated for this. In all four people left the department, all from NMR and MRI; they were Raymond Andrew, Bill Moore, Peter Allen, and Bill Derbyshire. Raymond Andrew had been negotiating initially with Tom Scott at Gainsville in Florida, but sadly he died at a relatively young age. Raymond was later offered a position there by Tom's wife, Kate Scott, as a physicist concerned directly with radiological aspects of MRI. Bill Moore was offered a position at the Brigham and Women's Hospital, part of the Harvard Medical School, and accepted this as a professor. He took the job to set up and head the new imaging group there. But after roughly six months working in the States he had a major heart attack one day while playing squash and eventually died. Peter Allen was, at the time, I believe, a Senior Lecturer at Nottingham, and he took a position at the University of Alberta. He set up and headed a very successful imaging centre there, now known as the Peter S. Allen MRI Research Centre. As far as I am aware. he followed his chosen research interest in MRI, but is now retired. Bill Derbyshire took a post at the Sunderland Polytechnic as a Senior Lecturer and kept that position for a year or so, but was eventually recruited to the Rank Hovis Company. Rank Hovis is a company that deals in flour refining and processing. Bill stayed in that job for several years and is now retired. All four of the voluntary leavers were concerned with NMR and MRI, and they all left within a six-month window during the period 1983–4. Of course, with their departure, all research in their particular areas at Nottingham ceased. In some cases, research students were transferred over to similar work at Nottingham. But in other cases the postdoctoral students simply left Nottingham and rejoined whichever group the leaders were now in but in their new institutions. The result of these moves created a considerable vacuum at Nottingham, but the important thing from my point of view was that all the infighting and intrigue that had gone on over the previous three or four years stopped. Suddenly we were in a completely free environment, where there were no concerns other than to concentrate on the imaging technique that we were developing in a completely uninterrupted and unabated manner, and this is what we did. In fact, I have previously described the ten-year era that followed from 1984 to the early 1990s as being the Golden Years as far as I was concerned, the golden years in our development of MRI, marred only by my mother's death in 1984, aged 78 years.

In the late 1970s and early 1980s, research students of my own group—namely, Ian Pykett and Richard Rzedzian—together with others in the group, had been very keen to press on to make a big effort to improve the quality of our images (see Figure 28). The question arose which way we should go, because, on the one hand, we had this advanced rapid imaging technique, which had not been experimented with very much and, on the other hand, we were able to produce line scan images of whole bodies but rather slowly. So a decision was made in the mid-1980s to concentrate all our efforts on EPI.[†]

This was done with the full participation of everyone in the group, and as a result we started to make considerable progress. There was, for example, work continuing on the theory and applications of EPI but also on the problems of generating the rather large gradients that were necessary and improvements in the receiver systems, the coil system, and so on. As a result we were able to produce images of relatively small objects, because the coil system that we used then was relatively small.

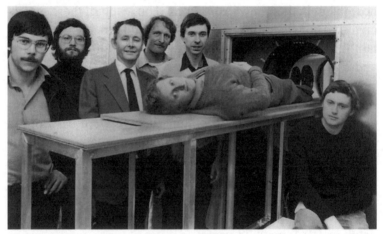

Figure 28 My imaging group, *c.*1980. From left to right: Peter Morris, Ian Pykett, Peter Mansfield, Volker Bangert, Richard Rzedzian, Barry Hill (lying down), and Roger Ordidge.

[†] See also Elementary Principles of MRI in the Appendix for more technical details of a range of imaging methods, including EPI and some commonly used variables and parameters in MRI...

Winston-Salem

In 1981, a bright Nottingham graduate in my research group, Roger Ordidge, had been working on the development of a small-scale gradient coil arrangement for imaging of objects using EPI. He had already produced some rather nice single shot images of red peppers and transverse slice images through his hand and fingers, since the receiver coil diameter was around 10 cm. We wondered if we could take some images of a small live rabbit. In fact he was able to produce a series of thin slice snapshot images through the thorax of this rabbit in which one could clearly see the beating heart. A movie loop was recorded and presented by Roger at the International Conference held at the Bowman-Gray School of Medicine in Winston-Salem, North Carolina, in spring 1981 (see Figure 29). Roger's talk was a show-stopper and was, in fact, the world's first MRI movie of a living animal. I spoke immediately after on theoretical aspects of EPI and resolution, but I must say that Roger's talk was a hard act to follow.

Later on we started imaging small children. Then with further improvements in receiver coil and gradient coil design; we were able to put ourselves in the machine to carry out whole body trials. Eventually we were able to attract for imaging numbers of volunteer patients with various problems and diseases.

One of the major areas that we concentrated on was cardiac imaging in children. We looked at cyanotic children in collaboration with a paediatric radiologist, Dr Alan Chrispin. We also looked at adults with various cardiac problems, myopathies and so on. As a result we began to produce quite nice images for which eventually, as the 1980s ticked by, the quality became more and more acceptable to our clinical colleagues who were helping with image interpretation.

In 1991 Dr Penny Gowland joined the group and began to make significant improvements in our efforts in paediatric

A. David Hoult   B. Roger Ordidge   C. Peter Hanley   D. Ferdi Buonanno
E. Jim Hutchison   F. Peter Mansfield   G. Leon Partain   H. Ergin Atala
I. Brain Worthington   J. Waldo Hinshaw   K. John Gore   L. George Radda
M. Larry Crooks   N. Graeme Bydder   O. Tom Budinger   P. Paul Lauterbur
Q. Rich Saunders   R. Frank Smith   S. Ian Young   T. Ray Damadian
U. Nolan Karstaedt   V. Paul Bottomley   W. Bill Edelstein   X. André Luiten
Y. Bill Moore

Figure 29 Conference members at the Bowman-Gray School of Medicine, 1981.

imaging, and in particular with studies of babies' growth and also in studies of fetal growth *in utero*. This important work continues today and has led to widespread recognition both within the university with her elevation to professor, and in the wider community.

Sir Rex Richards and the Oxford Debacle

In 1985–6, after Raymond Andrew's group and Bill Moore's group had left Nottingham and following my return from a series of recruitment interviews that I had attended in the States, I received an invitation to visit Oxford University.

On going to Oxford I met Rex Richards the former Vice Chancellor of Oxford University and a Professor of Chemistry who had initiated the NMR effort at Oxford. He told me that he thought that it would be possible to get me a professorship in Oxford. The professorship would be a professor *ad hominem* in the university. He asked me whether or not I would be interested in pursuing this. I said at the time that I was interested to explore the possibilities. Part of the deal would be to move my entire research facility at the University of Nottingham down to Oxford. There I would re-establish all my research facilities in a new building, which would be funded mainly by a £200,000 gift from Oxford Instruments. The gift would come from a charitable fund created by Sir Martin Wood, founder of Oxford Instruments. I should also add that the Oxford proposal, and in particular the proposal of Sir Martin Wood, was to build a laboratory for the MRI work. The intention was also to include the work of Sir George Radda, who was at the time working on phosphorous spectroscopic imaging.

At a subsequent meeting held with Rex Richards and Lord Dainton, a former Vice Chancellor of Nottingham University, it was decided to approach the University of Oxford with a view to creating a professorship *ad hominem* in the Physics

Department but with additional interests in the Department of Radiology.

It was something of a revelation for me to learn that all professors in the University of Oxford were paid the same salary and that that salary was at the time about £20,000 per annum. An additional shock came when I learnt that all professors in the university were not required to teach undergraduate students. It was explained to me that all undergraduate students were taught by tutors in their respective colleges and professors were not allowed to teach in the colleges. That was something I would have missed. At the time this created something of an impasse for me, because by that time I had received a salary increase, and I was being paid something like £23,000 per annum, a couple of thousand per year more than I was being offered at Oxford. In order to attempt to compensate the difference, it was suggested that the post in the Department of Radiology could supplement my income, so bringing it close to the salary I was currently getting at the University of Nottingham. I should say that this discussion went on for about two years, with no proper resolution, and I became tired of visiting Oxford and really getting nowhere in the discussions. So in the end I decided that I really did not want a professorship *ad hominem* or otherwise at the University of Oxford. It seemed to me to be politically too difficult, so I decided to stay on at the University of Nottingham. It was about this time that the NRDC received its first royalty payment from a settlement with the Technicare Corporation and subsequently from an agreed upfront payment from General Electric (GE).

A German Colleague

It was during the mid-1980s that my group was joined by a visitor from Germany, Michael Stehling. He had a degree in Medicine but he also had a degree in Physics, so he came

The Golden Years

well prepared, initially as a one-year visitor, but he decided towards the end of his stay that he wanted to remain at Nottingham to take a Ph.D. in Physics. So I took him on, and during the four years that he worked with us the quality of our images improved enormously. We were able to produce many papers covering particular aspects of a range of medical situations—for example, problems in the liver, cardiac problems, thoracic problems, and abdominal cases. As a result we were able in many cases to set the pace for high speed imaging, because virtually all the data that we produced in those days were in the form of snapshot images taking a maximum of around 50 milliseconds to produce. In some cases, particularly where there was involuntary abdominal motion or in cardiac imaging, we were able to take a series of images to produce a movie sequence, which often was extremely helpful in pinpointing the diagnosis that was made eventually.

These then truly were the Golden Years, as far as I was concerned, in the development of high-speed imaging and real-time imaging.

Of course, Peter Morris had left Nottingham in the early 1980s to work initially at the Medical Research Laboratory at Mill Hill and then after a few years he took up a lectureship in the Biochemistry group in Cambridge. Richard Rzedzian and Ian Pykett had also left and taken up positions in the United States. During that early period they also decided to set up a company, ostensibly to build MRI machines. They had received an invitation for funding from an American, Professor Haar at the Harvard Business School. Within a year or so they had managed to build an MRI machine capable of performing EPI experiments, and their plan was to interest hospitals in this machine. I think the original idea was to build these machines in numbers, but they had this one machine that eventually they got installed in the Massachusetts General Hospital in Boston. Although they were performing EPI experiments, I remember well the occasion when Ian presented a paper at a meeting in the States, where he insisted that he was not doing EPI experiments but something quite different. I was in the audience and questioned him about EPI, as I was convinced that that was exactly what he was doing. Later, during a visit to Japan in 1990, I recall speaking to Richard Rzedzian, who was also speaking at the Japanese Radiological Society meeting, and he admitted to me that strictly speaking their machine was indeed producing EPI data. In fact their machine was an improved variant of EPI. Later in 1997 the Rank Prize Committee awarded the Rank Prize jointly to Mansfield, Pykett, and Rzedzian in recognition of significant contributions to the development and commercial exploitation of EPI, not least in the development of novel pulse sequences. The Rank citation uses Instascan, the nomenclature of their company Advance Nuclear Magnetic Resonance (ANMR), for the enhanced EPI method. This involves a novel design based on rapid encoding pulses.

After protracted discussions, ANMR and GE entered into a collaboration agreement to develop jointly high-speed MRI for use on the GE MR systems. These systems were retro fitted by ANMR and installed at several institutions. Subsequently ANMR signed a long-term contract with GE. However, in the mid-1990s GE decided to implement EPI using a different configuration. In the meantime the machine at the Massachusetts General Hospital operating at 2.0 Tesla was used to produce a number of interesting medical imaging results, including the first demonstration of functional brain imaging using functional magnetic resonance imaging (fMRI). This application is now the principal use of EPI and its derivatives. It was also used to assess the effect of electric fields on patients when the magnetic gradients in the machine were turned up to relatively high levels. This work was fundamental to the widespread adoption beyond Nottingham of

The Golden Years

EPI and its derivatives, initially in the United States and elsewhere.

Back in the early 1980s, before Ian and Richard had left Nottingham, I was approached by the babycare company Johnson and Johnson, which had recently got into the imaging business. It had acquired a subsidiary company called Technicare, formerly Ohio Nuclear. In the early days it was quite interested in my type of imaging. In particular, it wanted to get hold of some gradient coils that were able to produce sufficiently large magnetic gradients. It sent a delegation over to Nottingham to talk about some collaborative deal. Part of the deal would be that Richard and possibly Ian would be involved in the production of these gradient systems for Johnson and Johnson. Unfortunately the deal never matured into anything tangible, but it caught the interest of Ian and subsequently Richard, inducing them to go into the business themselves, not here in Britain but rather in the United States. Johnson and Johnson/Technicare recruited Ian to work in the United States on its (non-EPI) MRI systems. Technicare subsequently became an early commercial leader in MRI for a time. Technicare, however, never had any strong interest in EPI, much to Ian's disappointment. That is a key reason why Ian left Technicare to form his own company (ANMR) to find a way to get higher-quality ultra-high-speed MRI methods developed and commercialized. It was Ian, not Johnson and Johnson, who recruited Richard Rzedzian from Nottingham to work for ANMR in the United States. Andrew Maudsley was also an employee of ANMR for a short time.

It was around this time, when Raymond Andrew was still at Nottingham, that he came storming into my office one day. He considered that I had sufficient electronic capability in my group and complained that I was using the departmental electronics facilities, which he and his group drew upon most heavily. I answered that I thought the departmental facility was there for everyone but that it looked as though it was catering for only one of the NMR groups in the department. We were also in NMR and we had no such electronic support, apart from the specific expertise of Terry Baines, which was part of our grant application. I told Raymond on this particular occasion that I thought the arrangement that existed in the department was very unfair. He then said: 'Well you are not part of the NMR group.' I was completely taken aback by this, and replied: 'Well I thought you had recruited me in the first place to be part of the group.' 'Yes,' he said 'but you've chosen to go your own way. Therefore you can't depend on our efforts in electronics. You must make your own way.' That caused a further rift in what was a very delicate position at the time.

10

Brief Encounter with Technicare

Lost Opportunities?

Back in 1984, I myself was courted to go back to the States. In fact I was courted on two occasions. Initially I went at the invitation of Johnson and Johnson, because it was rather keen to get me over to head up a group at one of the American universities in order to collaborate with them on the production of an EPI system. That never transpired, but my family and I did visit the United States. We went to New York and met a number of people in the J & J organization. On this visit to the States Jean and I took our two daughters and we were fêted at various places.

I remember the main people who were concerned with our visit over there were the American Jack McConnell and his English counterpart, Dr Scott, who was actually a consultant of the company. As it happened, he lived on Peckham Rye, not far from where Jean had lived previously when she was a young girl. The American trip took a few days in New York, and thereafter we were asked where we wanted to visit in the States. We were flown out to the Grand Canyon and Las Vegas, where we spent a couple of days, ending up in San Francisco and Los Angeles. Finally we were flown back home to England.

Also in the early 1980s I was headhunted by a number of other organizations. There were offers to go for interviews at the University of Illinois at Urbana, the University of Alabama at Birmingham, and also at the University of Wisconsin at Madison. Again I visited these institutions with

Jean and my two daughters, and at two of the places that we visited there was considerable discussion about the High School arrangements for my elder daughter, Sarah. She was at that time nearly 16 years old, and there was the possibility of her entering some sort of college either in Urbana or down in Birmingham, depending on which position I accepted. In fact I received very attractive salary offers for all three of the university positions. In Alabama, the offer included substantial funding for new equipment to work with Professor Gerry Pohost, but it was coupled with the idea of some sort of consultancy with the Philips Company.

It was while in Birmingham, Alabama, that I had occasion to speak to Governor George Wallace. He offered me some words of inducement to accept the position offered to me by the university, pointing out that there would be significant state funding available to support my research activities.

In Urbana, the offer was made from a private clinic that was setting up an MRI machine. The job that I was offered would be an academic position in the University of Illinois, either in Physics or in some suitable department, where I would do normal types of lecturing but also work in this medical centre. During the interview I was introduced to a whole host of people in different departments and presumably their opinions were solicited. During the course of the interviews I met Professor Charlie Slichter of the Physics Department, and he invited me to give a talk at a Colloquium in the Physics Department at Urbana. It was there that I presented our work on movie images and ultra high-speed imaging, which left a strong and favourable impression.

The third offer that I received was based on a GE offer of support in the University of Wisconsin at Madison. All three offers were attractive, but I made the point of not accepting any one of them on the spot. When I left each of them, I said I would consider very carefully the offers made and contact the universities once I had returned home to Nottingham.

Shortly after returning to Nottingham I received a letter from the Department of Health, and I was informed that unexpectedly a sum of money had become available, which was offered to me to buy a new 0.5 Tesla super-conductive magnet. This seemed to me like a golden opportunity, and I seized it instantly and accepted the offer of a new magnet. In the meantime I decided that, attractive as the offers were from the States, my family and I would probably be better off staying where we were, rather than accepting one or other of these offers with no guarantee that things would work out in the long run. I wrote back to all three universities expressing my regret that, because of the relatively new and changed circumstances back at Nottingham, I felt unable to accept any of the offers.

While Raymond Andrew was still in the department I had received money from the MRC to build an additional annex, a small building at the end of one of the research corridors in the department. The annex was not at that time joined to the main building, and reaching it involved leaving the corridor and going out into the yard area to access the new annex. The reason for having the annex was because we were engaged more and more in the imaging of patients, and it became necessary to have a separate building with separate entrances and toilet facilities, so that patients could come in without having to pass right the way through the building by the workshops and laboratories to get to the imaging suite. That had been a problem, but we had managed to get funding from the MRC to build the annex. Later, after Andrew had left the department and when the funding became available to buy the 0.5 Tesla super-conductive magnet, further changes were made to the building and the gap between the end suite of research rooms and the small annex was filled in, creating a little more space between what had been Raymond Andrew's old laboratory and the new annex. But, of course, all Raymond Andrew's activities had ceased, and that end room, where he had worked, also became available and was incorporated into the suite of rooms for the 0.5 Tesla imaging machine.

Department of Health

Interest in our imaging work by the Department of Health had been growing over a number of years through the efforts of John Williams, a member of a small team at the department that included Norman Slarke and Gordon Hickson. At the time, John's principal concern was in CT imaging support and the oversight of development of suitable phantoms for evaluating CT machine performance. John became increasingly convinced of the virtues of EPI and was, I suspect, the instigator of the super-conductive magnet that was being offered to us.

Many years later, I learnt that John lived near Royston and knew Eddy and Pat Dudman, originally from Camberwell, through a common interest in vintage motorcycles. After my nephew Steven's marriage in Great Dunmow, Jean and I drove by Saffron Walden and dropped in to see Eddy and Pat. John Williams also joined us for lunch.

Active Magnetic Screening

At about the time of the grand exodus at Nottingham I took on a new postgraduate assistant, Dr Barry Chapman. He had graduated from a joint honours course in Physics and Maths at Nottingham somewhat earlier and had gone on to Birmingham to do postgraduate work and to gain his Ph.D. We were fortunate enough to attract him back to Nottingham.

One day after the 0.5 Tesla magnet had been installed, I spoke to Barry about my worry concerning the magnet. The problem was the potential quenching of the super-conductive magnet when pulsing the gradient field on and off very rapidly.

Brief Encounter with Technicare

I had taken advice from the magnet designers at Oxford Instruments, who were not sure about the quenching problem but advised that it was not a particularly good idea to change the magnetic gradient field very rapidly close to the super-conductive windings. There was in principle the possibility of tipping the balance and triggering a major magnet quench. This had worried me while we were waiting for the magnet to arrive, and I had been thinking very seriously about how I could get around this problem. The magnet arrived and was installed, early in 1978. Oxford Magnet Technology personnel came up to Nottingham to align the magnet. We were ready to put our gradient coil in and start experiments, but the problem of potentially quenching the magnet was something that was nagging at the back of my mind.

I had already started thinking seriously about the problem of the super-conductive magnet and the possible interaction of a gradient coil system. Some experimental work had been started by Martin Cawley, and he had made a number of coils, which were simply ordinary circular coils wrapped round copper tubes so that we could see the effect of applying a current or a magnetic field and the exclusion effects that were intrinsic with the presence of a copper tube. This was the approach that I had been thinking about for some time, but the whole thing had not really gelled, and there was very little progress on this.

Suddenly, as if by magic, the whole problem crystallized in my mind, and on this particular day I rushed down into the lab and with great enthusiasm I said to Barry Chapman, 'I think I have solved the problem'. I explained to him that my solution was to use a magnetically screened coil so that outside the screen there was zero magnetic field from the gradient coil and inside the inner coil that was being screened was the gradient coil required for the experiments. I believed that such a system could be designed and built, so I did some calculations. In the end, when I had a proper and more substantial idea, I moved on to theoretical work, which initially involved myself, but I later got Barry Chapman to look into the problem as well. Between us we produced a theoretical paper entitled 'Active Magnetic Screening of Gradient Coils in NMR Imaging' with results that were the basis of a patent that was sent in to the British Technology Group (BTG) for consideration.

That is how Martin Cawley initially became involved. At that time he was a first-year research student and was looking around for an interesting area with which to get involved. His work rapidly got taken over by others in the group who were more experienced and able to continue with the ideas and ultimately produce the coil system that was designed.

In the meantime, while all this activity was taking place, another member of my staff, a new chap called Robert Turner, had recently requested to join the group. He had a theoretical turn of mind and thought he knew a different way to calculate the effect of screening on the gradient coil. His method of calculation was based on the so-called annealing method, and he spent several weeks working away at this without success. This all happened in a matter of two or three months. Eventually he got into conversation with a theoretician in the department, Professor Roger Bowley, and between them they were able to produce a theoretically sound method of both designing and screening the magnetic field gradient. Unlike our earlier attempts, this was a robust theoretical treatment that they set to work to implement. This was all done in a matter of six months or so from our original submission of the patent details, and under the rules for filing patents you have one year in which to add or subtract any items from the original patent. At this point it was decided between the four of us that we should include an additional section on gradient coil design based on the new mathematical method introduced by Turner and Bowley. All this was added within the one year allowed, and the patent was filed. The idea of active magnetic screening was born, and within a year or so of the patent being filed and papers being written on the technique the ideas behind active magnet screening were adopted universally by all the major manufacturers.

Echo Volumar Imaging

Another area that I have not yet covered was the work that had been done on echo volumar imaging (EVI). Early on I had written a paper outlining the basic principles and ideas for experiments in EVI. After publication of the theory of EVI, no experimental work had been done until Michael Stehling joined the group. Towards the end of his research period, when he was thinking about writing up his thesis, he wanted to include in it a specific area in which he could make some contribution but that was not part of any current project. I suggested that, if he wished, he could try some EVI experiments. He looked at the theoretical side somewhat, but spent most of his time programming the spectrometer so that an EVI experiment could be done. The necessary data manipulation programme was developed to look at fairly crude EVI images, containing 32 × 32 pixels in one image with eight contiguous planes. This was essentially a filler project for Michael Stehling. The next student that came along was Paul Harvey. Now Paul had been with me for some time, and for his initial project he had spent the first year and a half of his Ph.D. working with me on the effects of electric fields produced by placing small coils over the wrist and passing large transient currents through the coil to induce an electric field effect in the wrist. When this work came to a conclusion, the question arose as to what Paul could do next. Because of the previous work in EVI, started by Michael Stehling, I suggested that he take the EVI experiments to the next level and write some software to enable implementation of EVI at the whole body scale. This was done, and a number of experiments were performed with eight contiguous slices each comprising 64 × 64 pixel images all observed simultaneously using EVI. With this technique a number of volunteers were imaged, and such things as the study of aspects of gut motility were undertaken. In addition, studies of swallowing and transit of water bolii and the observation of bladder function were carried out successfully.

These initial studies were carried out on volunteers, with the MRI system operating at 0.5 Tesla using the superconductive magnet. Later, when we acquired the 3 Tesla magnet, new EVI experiments were tried out with the assistance of Ron Coxon and Paul Glover. The initial studies in this area were simply to look at cross-sectional images and a series of slices through volunteer brains in order to see whether we could collect a decent set of clinically useful images, eight contiguous slice images in this case, in a signal shot experiment. These results were very encouraging, and we were able to get acceptable images. A paper was written and the work published in the *Journal of Computer Assisted Tomography* (*JCAT*). A further set of experiments was carried out with the assistance of Ron Coxon and Jonathan Hykin, another medico who joined the group as a Ph.D. student, but this time we were particularly interested in observing functional imaging effects. Experiments showed that it was possible to get useful functional images simultaneously from eight planes. These results were also published in *JCAT*. Later on we learnt that this work had won an editorial prize that was awarded to three of us for the work on functional imaging of the brain. The paper was published in 1995. after I had taken early retirement in 1993.

Working with David Guilfoyle in 1985, I tried out another idea related to EVI, a hybrid called chemical shift imaging. I published the theory behind this work first in 'Spatial Mapping of the Chemical Shift in NMR', in *Journal of*

Physics D, Applied Physics, 16 (1983), L235–8, and later in more detail in 'Spatial Mapping of the Chemical Shift in NMR', in *Magnetic Resonance in Medicine,* 1 (1984), 370–86. Three dimensional plots of liquids with different chemical shifts were also obtained. This work was overlooked by others, who republished similar work much later and renamed it. Unfortunately, this outrageous behaviour is endemic within the MRI community and, in my view, arises through a mistaken belief that nobody will notice the mainly young miscreants and relative newcomers to the topic, all overzealous in their desire to make their mark, come what may.

11

Patent Affairs at Nottingham

The NRDC

It was 1974 when I first interacted with the National Research Development Corporation (NRDC), a government quango or quasi National Government Organization. The principal players in the NRDC with whom I interacted over many years were Graham Blunt, Norman Davis, Ruth Sutherland, and John McCarton.

Waldo Hinshaw sent off his application for a patent to the NRDC early in 1974. A month or two later I sent off my application for grant of a patent on slice selection combined with imaging. The patent attorney dealing with Hinshaw's patent turned out to be the same guy who received my patent, and so it was decided by the NRDC, as a matter of principle, that it was not a good idea for the same patent attorney to handle both applications. It was at that point that my application was handed to Norman Davis, another patent attorney in the organization. John McCarton's position in the NRDC was that of Executive Officer, and he acted as an interface between the patent attorney and the patentee.

I remember John telling me that, prior to his being appointed at the NRDC, he had worked for five years in Germany, I believe at the Bruker Company in Karlsruhe, and he spoke German fluently. He knew that I had spent time in Heidelberg, so he spoke to me in German, testing me out, so to speak, and we exchanged a few thoughts. Sadly a few years later John developed a brain tumour; he underwent an operation, but unfortunately did not recover and tragically died.

I found myself talking more and more directly to Norman Davis on detail and technical matters. But later on a Senior Executive, Graham Blunt, got involved in our work, and for a short while there was a Junior Executive, a young lady, who took over patent matters but eventually left. She was later replaced by Ruth Sutherland, niece of the operatic diva Dame Joan Sutherland. Both Joan Sutherland and her niece are Australians. I understand that Joan Sutherland, now deceased, lived in Switzerland. Ruth Sutherland used to commute from Switzerland to London to carry out her duties with the erstwhile NRDC.

It was not until the mid-1980s, some eleven years after the initial filing of our patent in 1974, that matters seemed to start moving on the revenue income front. The situation began to move when a decision was taken at Technicare, at the highest level in the company, that there was not sufficient business for them in the MRI field and that the company would be closed down, after five years in MRI. However, during the latter part of its trial period, Technicare had begun to make some MRI machines, and as a result outstanding royalties were owed to the NRDC. Because of the collapse of the company, an agreement between the NRDC and Technicare meant that some of Technicare's equipment was made available as a *quid pro quo* settlement to NRDC. On the announcement that the company was closing down, the NRDC moved to discuss the royalty situation, and Technicare's response was basically that, rather than give money, it would instead give equipment. As a result of this 'equipment in kind' arrangement, Technicare passed over to the NRDC a number of items of MRI equipment and a standard NMR console.

The NRDC's response was to find people within Britain who could usefully use this equipment. In one case, an NMR console was involved. This got shipped off to someone in Oxford, who was able to use the console for work being

Patent Affairs at Nottingham   157

carried out in drug development. As it transpired, the University of Nottingham, which was the major patent holder in MRI, was not considered as a suitable home for any of this equipment. When I heard about these arrangements, I immediately complained to the NRDC that the items that had been agreed in lieu of royalties belonged to, among others, the inventors of MRI and were not to be given away to arbitrary groups as the NRDC saw fit. This caused a bit of a problem, but eventually the NRDC conceded that there was financial value in these objects. This meant that the equipment had to be valued. Sums were associated with each piece of equipment, and the notional money was divided among the various parties, with the NRDC paying off the inventor's share. As a result of this a relatively small sum of money was made available, and the University of Nottingham received, in effect, one of its first royalty payments for MRI.

In the first instance the sum involved a few tens of thousands of pounds. I don't remember the exact sum, but it set the scene for future royalties and made our position absolutely clear that deals of the sort that the NRDC had attempted to try on were not acceptable. As it transpired, Technicare had tried to avoid paying any royalty or anything in kind, but there had been a legal case heard and as a result of this Technicare was forced to make some recompense to the NRDC. This was the turning point in the NRDC's fortunes, because very shortly afterwards GE, which had been watching what was going on in the litigation field, decided to settle with the NRDC by paying a lump sum royalty on machines that it had already built in addition to a substantial advance payment for the proportion of machines that it expected to build in the ensuing years.

In 1986 a very large down payment arrived with the NRDC, and this was the start of the substantial income that began to appear from the industrial companies, many of whom had only recently joined in the manufacturing of MRI equipment.

Shortly after this settlement was accepted by the NRDC, we filed a patent with the NRDC on active magnetic screening. This was an immediate success and was taken up almost instantaneously by all the MRI equipment manufacturers, including GE. However some six months after the initial filing of the patent, I went to the United States to attend a Society of Magnetic Resonance in Medicines (SMRM) Conference meeting and heard a paper presented by Peter Roemer from GE.

In his talk Roemer claimed to have invented active magnetic screening. In private conversations with one of his co-authors, Bill Edelstein, I was told that Bill had seen the paper that was sent by me to the *Journal of Magnetic Resonance* for publication. The paper was on active magnetic screening. Bill Edelstein apparently had either refereed this paper or at least had had sight of it before publication. But, in any event, Roemer had submitted his work for a patent something like six months after we had filed our patent in Britain, and so there was wrangling that went on for some considerable time. The end point in all this was that ownership of the patent on magnetic screening was considered in the European Patent Courts. We had long and protracted meetings and discussions where I was required to appear before tribunals, and the Americans were also required to be present at these meetings. Finally it was decided that we were the inventors, and this then unleashed a flood of back payments that were owed to the NRDC for the magnetic screening patent.

On another occasion I travelled to Washington with Graham Blunt, Ruth Sutherland, and Norman Davis to appear at a court hearing on our basic patent, which had been brought by one of the companies. I was required to give my version of events, and, ultimately, my explanation was accepted by the court. We left on that day, having won the case. When we got back to Nottingham, I spoke to the Business Manager, Derek Davis, and pointed out to him that not only had we invented the original ideas but we were also required to defend our ideas and furthermore we were involved in the process of giving talks and spreading the word so that others could pick up these ideas. My question to him was what was the University of Nottingham doing? It seemed to have done damn all in the whole process. I was doing the lion's share, so why should the university get half of any royalty income, particularly as I had never been party to any agreement with the University of Nottingham to share royalties with it.

Shortly afterwards I went to see the Vice Chancellor, Basil Weedon. He had not been around when the original patents had been filed, and his attitude was that any income arising from intellectual property and coming to the University of Nottingham belonged to the university. Of secondary importance were the guys who had created this intellectual property. He was of the view that it was the university's right to expect some income from inventions made on campus. I pointed out to him that I had never been party to any agreement on this and basically I disagreed with him, since I seemed to be doing all the work and the university did little except pay my salary.

Of course I also pointed out to him that all the funding that we had received over the years to do the research leading to MRI had not come from the university at all. It had all come from the MRC. I pressed these points upon the Vice Chancellor, but I could see that he was in a difficult position, because many new people who had joined the university more recently had been required to sign a document effectively getting them to agree to include the university in any revenue-sharing agreement if they filed patents on any ideas while they were in post. So the relatively small payments that were paid at the end of the 1980s, from 1986 or thereabouts, set in motion a major problem at the University of Nottingham between myself, on the one hand, and Basil Weedon, on the other, with the help of Derek Davis, the

Business Manager. As time went by the Business Manager's views were taken on board and accepted more and more by the university.

I had said previously to Derek and also subsequently to the Vice Chancellor that I had never signed such a document because this was not the practice when I had first joined the university. The situation when I joined was that one was invited to share revenue, but that, if one declined to do so, as we had done for our first patent, then that was accepted and that would normally have been the end of the matter. Of course on many of the subsequent patents I had been asked repeatedly by the Business Manager if I wished to participate in the university's revenue-sharing scheme and when I refused to do so the university ignored my replies.

In the end it became necessary for me to threaten the university with legal action, which I set in motion. In the event I engaged a fellow by the name of Jeremy Scholls from Wells and Hind, from a local solicitors in Nottingham that deals with intellectual property. He was a very sharp fellow indeed and had been involved in patent work previously. After discussions with him—rather costly discussions for me, I might add—a document was produced that set out the position as we—that is to say, the solicitor and I—saw it. On the basis of this I approached the university and the Business Manager. We had a showdown to the effect that, if the university did not agree, then legal action would follow. With this background I went along later to have a *tête-à-tête* with Basil Weedon, and he came up with a revenue-sharing agreement that he and the Business Manager had devised together. At my meeting with the Vice Chancellor, he gave me a copy of this agreement, but it had a maximum income, and there was no mention or discussion in the document as to what would happen to funds if they exceeded the maximum threshold that was mentioned. I raised this with the Vice Chancellor and said that I could not possibly sign the present document, but I would go away and, with the help of my own lawyer, would construct a new document and suggest that this be the one that the university accepted. I then went away and informed Jeremy of events. Together we constructed a different legal document forming the basis of an alternative revenue-sharing agreement. In this we introduced some really quite significant changes to the university's proposals.

I went back to the Vice Chancellor and proposed that I would sign a modified document, along the lines of our alternative revenue-sharing document. This took the royalty income up to a certain level and divided it in favour of the inventors. If income exceeded that level, royalties would be divided equally between the inventors, on the one hand, and the university, on the other. In my proposal there was no limit to the income level, whereas in the university's document there was a cap on the income, and anything above the cap was not discussed.

To cut a long story short, this document was produced and shown to the university, and, after discussions with the Business Manager, Basil Weedon readily agreed. The documents were signed, and the scene was set for all subsequent inventors in the university. Not only the inventors of MRI but also others in other departments were bound by the agreement that Jeremy and I had drafted.

Although I was not overly happy to sign it, I did eventually sign the document because at that point there were funds coming to the university that would otherwise have been put into escrow. I felt that the money should be distributed as soon as possible, so that we could all get the benefit, including the university. At the end of the day we did agree, the money was released and dispersed, and that is the situation that exists today.

Prior to this agreement and after the Technicare settlement, there was a brief period of about six months when the first sizeable payment of royalties came along. This came from the basic patent that involved Peter Grannell, Allen Garroway, and myself. There had been no agreement on that very first patent, and, as had been customary in those days, I was asked whether I wished to include the university's participation in the funding or not. In the event I had said I was not willing to include the university's participation. So that single patent had got through the net, and the money that came from the NRDC all of it ultimately was split between the three inventors. But as I said earlier, following this payment there was discussion about the new agreement. That was accepted ultimately, so that all subsequent income from NRDC patents was split according to the new rules.

Shortly after this incident, the NRDC decided that it wanted to honour me with a special gift, and in addition it wished to host a reception party in order to make the presentation. This was to be held at the University of Nottingham. So a suitably inscribed solid silver plate was presented to me at a reception dinner, and of course the Vice Chancellor, Basil Weedon, was present. I do not think he knew at the time that there were substantial payments that had been agreed by GE and one of the reasons why I had wanted to get the royalty business settled was so that the next payment, which was going to be very substantial, would be paid without any of the money being held in escrow.

BTG Formation

In about 1990 the National Research Development Corporation (NRDC) changed its name to the British Technology Group (BTG). This was no longer a government quango but in fact an independent commercial company that exploited patents and ideas arising from universities. The difference, so far as the universities were concerned, was that there was no longer any obligation to make sure that university employees filed patents through the BTG, although that seemed at the time the best approach for us. However, as time went by and most of the valuable MRI intellectual property held by BTG had expired, the group decided that it would relinquish all further interest in MRI. Instead it decided to concentrate its interests in pharmaceutical patents.

This change required an Act of Parliament. At the time, a young junior minister in the opposition Labour Party, a certain Gordon Brown, sought my views on the proposed change. Even then, I could see dangers to the interests of inventors and communicated my concerns to Gordon Brown, who then raised the matter in Parliament. His comments on my concerns are recorded in Hansard.

12

A New Vice Chancellor

The Campbells are Coming

In 1987 I was invited down to Oxford for a British Association (BA) meeting. Among the topics that were being considered at the BA meeting was the question of investment in research and development. One of the sessions was chaired by somebody from the NRDC. I had been invited to present a paper at this meeting talking about the process of invention and how the NRDC had handled the patent position and so on. Unbeknownst to me, within the audience was a Professor Colin Campbell. He did not make himself known to me at the meeting, but later I learnt that he was the heir apparent to the Vice Chancellorship at Nottingham. When Basil Weedon retired, around the end of 1986, Professor Campbell was appointed as the new Vice Chancellor at Nottingham. It was several months afterwards that I was invited to a meeting in the new Vice Chancellor's office, and the matter of patents was broached. It was at that point that Colin Campbell informed me and others at the meeting that he had been at this meeting in Oxford, and, in his words, had been 'very impressed with my performance'.

The meeting eventually got around to talking about the MRI work that we were doing and the Vice Chancellor asked me if I felt that the university had been doing enough to help out. My response was that I had had considerable problems in the past and I was really on the verge of trying to expand our activities but I seemed to be thwarted at every stage. The Vice Chancellor was very concerned about this. He said at the meeting that he thought the university should be doing a lot more to help out. He then invited me to submit plans and proposals for our new venture—namely, the new Magnetic Resonance (MR) Centre.

It was obvious that we could not expand further in the Physics Department. In any event, such an expansion was not conveniently placed. Shortly afterwards the Head of the Department of Works came down to see me, and we started looking around the university campus to see where a suitable place could be found to build the MR Centre. After spending some considerable time trudging around looking at various sites, we settled eventually on a site just off the main visitors' car park. This was agreed by all, but of course the university then needed to go to the local authority to get planning permission. As it happened, planning permission was granted pretty well straight away, because most of the university is a green site and there was no real problem of interference with other people and other buildings. We obtained the permission and shortly after that the university started to look seriously at architects and eventually builders to start the construction. Now this all took place over the two years or so after the Vice Chancellor was installed in his new post, leading up to 1990. The building was actually completed in September 1991 and then subsequently was opened officially by the then Secretary of State for Health, Virginia Bottomley, on 10 May 1993.

Just before we started to move into the building, the Vice Chancellor asked me whether I wanted the new MR Centre to be an independent building and department in its own right or whether I wished to continue to be part of the Department of Physics. My instant reply was that I would prefer to continue to be part of the Department of Physics, because many of the research students that hopefully would join us in the centre would be recruited from the Physics Department. However, I had been having some considerable problems with the Physics Department at the time. This was because there were others in the main body of Physics who were profligate spenders of money and as a result the Physics Department ran into severe debt within the university. It created a major problem, particularly for those who had been thrifty with the departmental grants and had managed to accumulate money. I was very concerned and of course the other contentious issue was the fact that the university started to support the Physics Department with some of its royalty income arising from MRI. On several occasions I had to go cap in hand to the Physics Department to get some of the money that had actually been paid to them by the university on the back of our royalty income. It was, therefore, an unsatisfactory situation as far as I was concerned. In those early days I began to think seriously about whether I had done the right thing by so readily agreeing to be part of the Physics Department. Had we remained separate and been a completely independent department, my suspicion is that we would have fared as well if not better.

Shortly after we had installed ourselves in the new MR Centre I saw Basil Weedon in the car park outside. He had returned briefly to the university for a business meeting. I remember waving to him and asking him how he was. It was with great surprise and dismay that I learnt later around 1992–3 that he had died.

Colin Campbell, in my view, made a number of important changes to the university, particularly in the way it was run. There was considerable murmuring against his ways and methods, but from my point of view he was extremely helpful at every turn. His support was the principal reason why we now have an MR Centre. However, I should say that, in pulling together the various financial agencies, I was able to persuade the Department of Health as well as the BTG to contribute significantly to the costs of the new centre. The university also contributed but to a lesser extent. Colin Campebell suggested involving the Prime Minister, Margaret Thatcher, to stamp her *imprimatur* on the arrangement. This was done with a small celebratory meeting in Downing Street around 1990 (see Figure 30).

A Knighthood

In around October 1992, to my great surprise and astonishment, I received a letter in the post from the Office of the then Prime Minister, John Major, to the effect that he was minded to recommend me for a Knighthood in the Queen's New Year's Honours List and basically would I accept it? In the letter I was told that this information should be kept strictly confidential until the New Year's Honours List was announced. This was extremely difficult to do, because I felt

Figure 30 Celebration of the NRDC/Nottingham accord at 10 Downing Street. From left to right: Ian Harvey, Colin Barker, Mrs Thatcher, Colin Campbell, Peter Mansfield, and Peter Morris, *c*.1990.

A New Vice Chancellor 169

it was necessary to inform the family of the occasion, and of course it was difficult to guarantee that the secret would be kept. Nevertheless Jean and I did our best to keep the news under wraps and decided to celebrate the announcement of the award at a New Year's party held in one of the hotels in Nottingham.

As it transpired, there were two other people from Nottingham, one in the Queen's Medical Centre, Professor Hull, and Professor Smith, who had retired but had been Professor of Law in the University. So all three of us were to receive Knighthoods, and the newspaper in Nottingham, the *Evening Post*, made a big splash about this. Jean and I invited the other two Knights to come to our house, and there in our lounge the photographer came and took some nice pictures of all three of us together that eventually appeared in the *Nottingham Evening Post* (see Figure 31).

The Knighthood itself was conferred by the Queen at Buckingham Palace later in the new year (see Figure 32).

These were heady days so far as I was concerned and I began to think seriously about whether it would be sensible to take early retirement so that I could concentrate any spare effort that I might have on my current research topics at the time. I also suspect that I was strongly influenced by the death of my eldest brother Connie (Conrad William Mansfield), who had died in 1993, aged 68. Eventually I decided to take early retirement in 1994 at the age of 61 and made my position known to the Vice Chancellor, who was understandably disappointed. But, as I said to him, it was a way of spending more time on my research. So it was that I retired from the university as a full-time professor (see Figure 33).

In the meantime, Dr Roger Ordidge had rejoined the Physics Department. He had spent time working for Bruker, but eventually decided that he would quite like to return to academia. I was able to persuade the university authorities that we could use an additional person in the MR Centre to

Figure 31 Professors Hill and Smith photographed with me by the *Nottingham Evening Post* at my home, 1993.

assist with the development of MRI work, and Roger was appointed. I believe he came back to the university as a lecturer.

Roger subsequently gave up his post at the University of Nottingham and took a Professorship in the United States. He spent a few years working in the States on MRI and eventually returned to Britain as a professor where he was appointed at University College in London.

Shortly before the new MR Centre came fully on stream the university decided to appoint a Reader to assist me in my work in the Centre, and the position was advertised.

A New Vice Chancellor

Figure 32 Family celebration of the knighthood at Buckingham Palace, 1993. From left to right: Sarah, Peter, Jean, and Gillian.

Figure 33 My retirement party, 1993. From left to right: Allen Garroway, Jean, Peter, Mary Garroway.

My former student Peter Morris was an applicant. Peter at that time was a Lecturer in Molecular Biology in Cambridge and he came for the interview, together with others. In the course of the afternoon of interviews Peter Morris was offered the Readership. However, at the interview he decided that he really was not interested in a Readership but wanted a Professorship at Nottingham. This created a considerable problem on the spur of the moment for the Vice Chancellor, but he eventually resolved the position and, after speaking to me, felt that he was able to change the terms of the advertisement, and Peter was appointed Professor of Physics.

Following Peter Morris's appointment, he was involved with me in the very early stages of planning the new building. We had already seen draft plans of the building, but Peter requested that additional space be made available in the form of a biochemistry laboratory.

One of my growing concerns in those early days after we had moved in to the new building was the question of the safety of MRI. A problem that seemed to be increasing as time went by was the question of acoustic noise and the very high levels associated with gradient switching when performing very short scan EPI sequences in less than 50 milliseconds, for example. The noise levels registered on an acoustic sound meter were quite often in excess of 120 and sometimes reached 130 decibels. As we really wanted to do even shorter and faster sequences, the noise level was becoming a danger, not only to the operators but also and especially to potential patients who might be imaged. I therefore set about initiating some experiments together with Paul Glover and a shared research student, Joanna Beaumont. She was co-supervised by Richard Bowtell.

This research was our first foray into the world of acoustics, and in those early days we looked at the acoustic response of various plates of material but still did not have a clear idea as to how one should proceed in order to reduce

A New Vice Chancellor 173

the acoustic noise level. One idea was to take a particular plate size and divide it into three narrower plates, so that the response overall was just three times the response of a single plate. The single narrow plate response could be made to be fairly small. When Joanna Beaumont came to write up her thesis, the acoustic work made one chapter and was a relatively small but valuable part of her work overall.

Fluid Flow in Sandstone Bore Cores

During the period from 1994 to around 2002 I had three other research students, but these were in a non-medical imaging area. The first of these was an Iraqi student, Bashar Issa. The second student that followed on was an Omani student called Mohammed Al-Mugheiry and the third and final student was a French fellow, Martin Bencsik, who had come from the laboratory of Jacque Chambron in Strasbourg. Prior to 1993 I had developed an interest in non-medical imaging and had started work fairly early on with Richard Bowtell as my first student in this area. We worked on aspects concerned with the oil industry and other non-medical applications of MRI. This work was continued by Stephen Blackband as part of his Ph.D. studies, until he left and went first of all to the University of Hull and then later to the United States. As time progressed I developed an interest in imaging bore core materials obtained from the drilling process in the oil industry. One of the problems that had been identified was the measurement of some of the flow characteristics of water absorbed in and flowing through bore cores. The work on oil bore cores was funded by the British Petroleum (BP) Company at the time when Ken Packer was the Head of the Research Facility at BP.

During the early work that Bashar Issa carried out, we noticed that the flow images through certain sandstone bore cores, when repeated at the same flow rate, quite often showed marked differences in the flow pattern. Initially we attributed this simply to thermal noise but on repeating these data again at the same flow rate I came to the conclusion that the signal variation that we were seeing was much greater than the thermal noise one would expect in these images, and I started to think very seriously about the cause of this effect. It turned out later that the rather large apparent noise fluctuations that we were seeing were connected with what we called hydrodynamic noise. This was a time-independent statistical fluctuation in the arrangement of flow channels within the bore core. Mathematical analysis based on a statistical approach to the flow problem revealed that the flow variance was proportional to the mean flow rate, whereas mean flow rate itself was controlled by Darcy's Law. We named this unexpected linear behaviour of the flow variance *versus* the mean flow velocity the Mansfield–Issa equation. The Mansfield–Issa equation was tested on various bore core materials and found to hold in all these examples.

Mohammed Al-Mugheiry continued the work studying bore cores, looked in more detail at the theoretical consequences of the theory, and also tried to extend the theory for the Mansfield–Issa equation, which did show signs of a small deviation from linearity at much higher mean flow velocities.

Martin Bencsik initially arrived in my group for a one-year period, as he was on special leave of absence from Professor Chambron's laboratory in Strasbourg. But after a few months he realized that he liked the work in which he was involved and decided that he would like to stay and work for a Ph.D. full time. He managed to arrange to be released from his job in Strasbourg and stayed at Nottingham for three years. During that period he got deeply involved in the flow studies but from a slightly different point of view. Instead of looking at actual bore cores in order to test out the Mansfield–Issa equation in much more detail, Martin decided to look at assemblies of glass beads of different sizes in order to simulate, in a fairly clean and more accurate manner, the effects of a porous medium. He used these glass bead assemblies to study the flow characteristics of water as it passes through the bead beds. In all cases he was able to show that the water flow seemed to obey the Mansfield–Issa equation. The largest bead size that he tried was approximately 1 cm in diameter and even here he found that for a range of mean flow velocities the flow variance again obeyed the Mansfield–Issa equation.

Acoustic Control

In the meantime I had managed to persuade my son-in-law, Brett Haywood, to leave his work at the Electricity Board and come to work for my small company, General Magnetic. In addition, I suggested that he might wish to take up a part-time course at the University of Derby, studying for an electronics degree in the Electronics and Electrical Engineering Department. He already had his HNC and HND qualifications and so was accepted at the University of Derby and allowed to move straight into the second year of his undergraduate degree. In the two years that followed he designed and constructed apparatus for General Magnetic that also formed part of his experimental project for the University of Derby. The project was broadly based on the acoustic work that had been performed earlier with Paul Glover and Joanna Beaumont, but this time with the added concept of acoustic control that was included in the plate designs forming a gradient coil assembly.

While working at the MR Centre, Brett was able to pick up on the principles of NMR and particularly MRI from working occasionally with Martin Bencsik. During the course of his undergraduate work, Brett built up a number of electronic circuits that were able to produce signals in which the phase and amplitude of four independent channels could be accurately controlled and fed on to high power amplifiers for application to the acoustic control windings in the modified gradient plates. These were tested at 3 Tesla in the MR Centre. The control-winding concept seemed to work exceptionally well in individual plate pairs that were tested. In several examples we were able to achieve something like 50 decibels reduction in noise level but always at a particular frequency. The concept, though it worked well, was relatively narrow band and depended very much on the right choice of operating frequency in the plate.

This operating frequency seemed to be dependent on the natural resonance frequency of the plate and worked particularly well if one operated at or close to this natural resonance frequency. However, if one operated off the plate resonance peak, the noise reduction became less and was nearer to a 30 decibels reduction. Shortly after Brett graduated in Electronic Engineering I bumped into a former colleague of mine, Barry Chapman, who had returned to Britain and was living close to the university campus. I asked him what he was doing and it turned out that, among other things, he was writing books, but at that time was without a regular job. I therefore suggested that he might be interested to work for General Magnetic. He accepted the offer of a part-time position, which he continued until 2009.

Meanwhile, when Brett had finished his undergraduate degree, he decided that he would like to stay on to study for a part-time Ph.D. This was a four-year course to which I readily agreed. I took him on as a part-time Ph.D. student but concentrating on the acoustic work, since we had already shown that the principle of acoustic control was possible in single plate-pairs. The next question was whether we could apply this to an assembly of plates forming a gradient coil.

This work started in 2001 and finished in 2005, with the demonstration of a crude phantom image together with

A New Vice Chancellor

a slight reduction in acoustic noise of about 12 decibels. Thus the basic concept of producing a gradient for rapid imaging with an accompanying noise reduction had been shown to be possible in principle, although the quality of the data meant that it should perhaps be repeated on a future occasion. In fact during the following year a number of important modifications to the gradient coil system were made. In particular, it was found that the thickness of the plates themselves gave rise to slightly different resonance frequencies. So all the plates were shaved down to a common thickness and that meant that all the natural resonance frequencies of the plates were equal to within about ten or so Hertz. The experiments were repeated just over the Christmas period. Again it was not possible to do both the acoustic experiments and the imaging experiments at exactly the same time, because we did not have a sufficient number of amplifiers to drive the control winding while we were performing imaging. It was decided that the imaging experiment would be done first, followed by the acoustic experiments. The results that we achieved were much better. The image quality that was obtained after quite a lot of experimental work for optimization was really quite good and the imaging time was 10 milliseconds.

The acoustic experiments were done afterwards with the available power. But again the power available was just not enough to power the control winding for acoustic noise reduction at the operating gradient strength. It was decided, therefore, to do the acoustic work at half the gradient strength and rely on scaling of the data to get the noise that would have occurred had we been able to image simultaneously with the acoustic experiments. This time round we were able to achieve something like 40 decibels noise reduction effectively during the imaging process. The images were one shot but in order to improve the quality somewhat we decided to take two single shot images and co-add them to obtain an improvement in signal/noise ratio.

Philips, BTG, and Active Magnetic Screening

In 1986 Barry Chapman and I had published a paper entitled 'Active Magnetic Screening of Gradient Coils in NMR Imaging' in the *Journal of Magnetic Resonance*. We had earlier filed a provisional patent together with Roger Bowley and Bob Turner. This was an immediate success commercially and was taken up by the leading MRI equipment manufacturers, including Philips.

However, in the mid-1990s, Philips stopped paying royalties for active magnetic screening, even though it was using the concept in its MRI scanners. I complained to the newly formed BTG and was told subsequently that it had taken legal advice and had apparently been informed that it was unlikely that the BTG would win in any legal action.

I later went to see Colin Campbell, and he suggested that both he and I should meet Jaques Coumins, Philips's representative, in London. This we did, and I made the point that Philips was using our patent and violating the BTG agreement and should be paying royalties. Coumins said he would take up our complaint with the Philips hierarchy.

A considerable time elapsed, so I prompted Colin again. He wrote to Coumins, and eventually it was agreed that Coumins would meet me in Leiden at a forthcoming MRI meeting to which I had been invited. The meeting was to celebrate the installation of a Philips 7.0 Tesla scanner in the Radiology Department of the Leiden University Medical Centre. After I had given my talk, there was a reception at which a whole host of Philips people were present, including the Managing Director of the MRI division. I asked after Coumins, but he was not present. I learnt later that he had been transferred to a different department.

When I returned to Nottingham, I explained the position to Colin, and he said he would make further enquiries.

After several weeks I went again to see Colin. He thought it was worth one last try with a direct approach to Philips. A week or so after this the Vice Chancellor took retirement, and so far as I am concerned the matter remains unresolved. At the present time I estimate that Philips owes the parties to active magnetic screening around £5 million, of which approximately £1.5 million is currently owed to the inventors.

The reluctance of the BTG to pursue and prosecute our infringement case with the Philips Company, concerning its use of active magnetic screening, demonstrates, in my view, that with the passage of time, my fears about the privatization of the NRDC were well founded.

13

Nobel Prize Speculation

Just a Daydream

In 1993 the American Journal *Diagnostic Imaging* published an article by Philip Ward speculating on the Nobel Prize and canvassing opinion on who might be awarded the prize for the work on MRI. The journalist had phoned around and managed to obtain telephone interviews with some people on this side of the Atlantic, and in addition had gone to interview Paul Lauterbur, Raymond Damadian, and many other people in the States to obtain their views. I received a telephone call myself, but I declined to be interviewed and had no comment on the whole process. I really felt at the time that that was the sensible thing to do, since the whole question of who might or might not get the Nobel Prize could well have been harmed by wrongly placed comments. In the circumstances, therefore, I declined to make any comment. Subsequently my colleague Professor Brian Worthington told me that he had also been contacted by this American writer and he too had declined to make any comment. Eventually an article entitled 'Is there a Nobel Prize in MRI's future?' was published in the following issue of this free journal. The two- or three-page article contained a general history of the Nobel Prize and comments about who might or might not get the award, plus a whole list of people who had given comments, including several Americans, some Swiss people in MRI, and also some English people. There was no general consensus among these people as to the line-up of recipients. However, everybody seemed to be absolutely clear and certain about one person who should get the Nobel Prize and that was Paul Lauterbur. With the passage of time interest in the article eventually waned and in due course I virtually forgot about the whole episode.

I genuinely believed that, following this stir of interest and the fact that there had been no announcement, the whole business had been largely forgotten, not only by me but by the Nobel Prize Committee itself. It seemed that there was really no chance of obtaining a prize in the circumstances. In fact, I began to think that the intervention and public discussion that had occurred at the time of this article in *Diagnostic Imaging*, and the comments made by those interviewed, had severaly marred the situation.

I had taken early retirement and in the following years concentrated on the research topics in which I was interested. I was putting in a great deal of effort on the acoustic noise problem and that looked as though it would work, at least in single plate pairs. We had already demonstrated a considerable reduction in noise, but, as I said previously, this occurred at a particular frequency close to the plate resonance frequency.

As the new millennium approached, Jean and I decided that we would celebrate the occasion by going to Stapleford Park Country Hotel, taking the family with us. This we did on the New Year millennium occasion, and we enjoyed the splendours of the millennium party and all the excitement, fireworks, and so on that went with it. It was a most enjoyable occasion.

Almost one year following the millennium, in October 2000, I was diagnosed with prostate cancer. Following the diagnosis, arrangements were made for me to have a radical prostatectomy in the following January. The operation went well, and I was informed afterwards that the prognosis was rather good. However, the few weeks following the operation were a somewhat awkward period, because of drainage tubes and urine collection bags that were uncomfortable and at times somewhat embarrassing. Brett's first-degree presentation was made in February 2001, and I was asked to be present, an invitation that I gladly accepted. Of course I still had the recovery tubes in me and it was slightly uncomfortable, but nevertheless we got through the afternoon and back home again quite safely and satisfactorily.

The Telephone Call

Our efforts in quiet gradient coil technology intensified over the following 2½ years. Then in 2003, on Monday, 5 October, at around nine o'clock in the morning to be precise, the telephone rang. At the time I was in the bathroom shaving and generally preparing myself to go to work. My wife called up and asked me to go quickly to the phone. Someone in the university wanted to talk to me about the Nobel Prize. I shouted down some comment like, 'Pull the other leg, it's got bells on'. I was just finishing off my ablutions and really did not appreciate any jokes at that time of day. But Jean replied, 'No, it's no joke, come down quickly.' So I went downstairs and someone from the university was on the phone saying that there had been a call from the Swedish Embassy and that this person would be calling back to give me the details of the Nobel Prize that had been awarded. So I went back upstairs to finish my preparations, and literally within ten minutes there was a second call, but this time it was from the Swedish Embassy and someone, I believe the Swedish Ambassador, said that Paul Lauterbur and I had been jointly awarded the Nobel Prize for Physiology or Medicine and many congratulations. I would be getting more instructions later on. I was in a dazed state of euphoria at the news, and it took some time for the words of the Ambassador to sink in. Within about half an hour of my receiving the message

I started to get calls on the telephone from friends and colleagues who had heard the news on the radio or television and were phoning in their congratulations.

Slowly the implications of the Nobel Prize began to sink in, and I started to worry about the enormous responsibility of such a burden. However, I soon decided that my approach would be to take the whole thing in my stride and handle it the best way that I could. A week or so later the Physics Department decided to throw an impromptu celebration, organized at the last minute in the University Staff Club. I did wonder about the venue, but for some reason they could not get the area downstairs, and we were crammed into a couple of small rooms in the upstairs of the Staff Club. I went along to this meeting and, just as I was entering the Staff Club, I saw Professor Rex Coupland ahead of me. He and all members of staff were going upstairs. The Vice Chancellor was also invited, and at an appropriate point I was asked to say a few words. Because the whole thing had been organized at such short notice, I gave an impromptu speech of the cuff. In such circumstances it is often quite difficult to remember or include all the people who have contributed to your success. On this particular occasion I felt that I had not done justice to all the various people that I know had been associated with me over the years and I found the whole business rather embarrassing. I really had not had enough time to think about things or prepare my speech. Sadly the two principal medical collaborators have since passed away. Professor Brian Worthington died of pancreatic cancer in December 2007. Professor Rex Coupland died following a major stroke in July 2008.

In the week following the telephone announcement of my Nobel Prize I received a package from the Nobel Institute in Stockholm containig details and guidelines for the preparation of the actual award ceremony to be held in Stockholm. I also heard that there had been a scurrilous attempt by

Raymond Damadian to denigrate both my position and Paul Lauterbur's with regard to receipt of the Nobel Prize. The announcement was made in a two-page advertisement that was funded and launched by Damadian in several newspapers both in the United States and in Britain. The British version appeared in *The Times* and purported to tell the story of MRI, its discovery and invention by Damadian, and why the Nobel Committee had made a terrible mistake in proposing me and Paul Lauterbur as recipients. Of course the news media pounced upon this, and I had a stream of telephone calls from reporters anxious to hear my side of the story. In all cases I refused to comment except to say that I thought that the article was an offensive attempt to blacken in the worst possible way the two of us and the work that we had both put into the development of MRI for the ultimate medical benefit of mankind. Of course, such a comment was not reported by the press. They were much more interested in tawdry comments and sniping accusations but were disappointed at my response of no comment.

When we read the detail in the information pack of the Nobel Prize Committee, it became apparent that they would only cover the costs of travel to Sweden for me and my wife. A number of other people were eligible to come but at their own cost. I decided that the sensible thing would be to limit the number of guests to our close family, namely my two daughters, their husbands, and the four children. However, I received a telephone call from Allen Garroway, who phoned in his congratulations and asked if he could come to the ceremony in Sweden. I explained the financial position to him, but he decided to come together with his wife at their own expense.

Gillian, Brett, and their two children, together with their nanny, travelled by aeroplane. Sarah preferred to come by train to Stockholm with her two girls and her husband, Mike. As soon as my party arrived in Stockholm, we were ushered to the VIP lounge and then through a back door, completely bypassing immigration and customs. However, our passports were taken, checked, and then returned to us. Eventually we arrived at a special gate on the land side, and waiting there were a couple of limousines. We all took our seats in the limousines and were driven to the Palace Hotel in Stockholm.

When we arrived at the hotel I noticed that there were police guards at the front of the hotel and I wondered whether there had been any problem with Damadian. So once we had been ensconced in our various apartments, I made enquiries and found that a guard had been placed at the front of the hotel because trouble had been expected. Apparently, so the story went, Raymond Damadian was planning to arrive in Sweden and intended to create problems for everyone. The police were there to prevent any difficulties. In the event no such problem occurred, and later in the week we heard, contrary to what we had been told previously, that Damadian had decided to organize a meeting in the United States on the same day as the Award Ceremony was taking place in Stockholm. Several people from Stockholm had been invited by Damadian They were people who disagreed with the Nobel Institute, and they went off to the States and held their Anti-Nobel or Ignoble Prize Meeting. Because of this we all soon forgot completely about Damadian and the problem that he had created. However, on the day of the Nobel Prize Lectures, a huge two-page advertisement appeared in the Swedish Newspaper the *Dagens Nyhetter*. Of course, I could not understand a word, but I rather assumed from the layout and diagrams that it was a Swedish translation of the two-page advertisement that had appeared in *The Times* in London a week or so earlier.

The Talks

When we had finally settled into our hotel room, we found a package awaiting us. In it were a number of invitations to various events, plus an itinerary for the week. Among the various activities in which we were expected to participate was, of course, the presentation of the lectures. It turned out that all the Nobel lectures were given on the same day, the Monday, but not in the same institute. The Physics talks were given in the Physics Department on the campus of the University of Stockholm, the medical talks given by me and Paul Lauterbur were held at the Karolinska Institute, and the Chemistry talks were given in the Chemistry Department. In some cases, because of the timing of events, there was a little overlap, but I noticed that the Physics talks were given at a slightly different time from the talks that we were going to present, and therefore Brett and I managed to slip away to hear the Physics Laureates give their talks. The whole process of moving around the campus and indeed moving around Stockholm was made easier by the fact that we had been assigned a limousine and driver. We also had a guide allocated to us at virtually all times. Our guide was a young lady from the Swedish Foreign Office. She could speak English exceptionally well. Our driver, Fritz, who was originally from Germany, spoke some English but mainly German and Swedish. That was fine, since it meant that I could communicate with him in German, and we got on quite well together during our week's stay.

On the Monday morning I was able to hear Leggett, Ginsburg, and Abrikosov give their separate lectures on the work for which they had been awarded the Nobel Prize in Physics. In the afternoon at the Karolinska I noticed that Paul Lauterbur's entourage included Waldo Hinshaw and Don Hollis. My talk followed Paul Lauterbur's. Because he had had a stroke sometime prior to receiving the Nobel Prize, he was not completely fit. Nevertheless he gave a one-hour talk, but on this occasion chose not to talk in detail about his work in MRI. Rather he spoke about the historical background to his work and basically his life story up to that point. Immediately after Paul had spoken I got up to give my one-hour talk. This was a more conventional presentation on MRI and my particular approach to it, with slides and diagrams. In many ways my talk, at least in style, was similar to the three Physics lectures that I had heard during the day.

Among the various events scheduled on our personal itineraries were a number of interviews carried out on a one-to-one basis. On the Wednesday there was a general discussion about certain topics, in which all the Nobel Prize winners in Medicine, Chemistry, and Physics sat around one big table and offered our own views about the topic being discussed. This discussion was chaired by a BBC interviewer, and the whole process lasted about an hour. Following the discussion Paul Lauterban and I remained seated and continued to chat (see Figure 34).

Prior to going to Stockholm I had been asked for my suit size. It turned out that all male Laureates were required to

Figure 34 Sharing an amusing tale with Paul Lauterbur, 2003.

wear white ties and tails for the prize-giving ceremony. On the Wednesday morning I was driven to a suit hire shop in Stockholm. There we tried on suits that had been previously ordered and any small adjustments were made.

On the Wednesday evening I had received an invitation for our entourage to visit a big student gathering and party that was being organized by the graduates and undergraduates of the university and we thought that it would be impolite not to go, so we made our way to the area where this event was taking place. Eventually we found our way into a large awning where a lot of the activity was taking place. We had a few drinks and stood for half an hour or so watching the various activities and dances before quietly slipping out, but at least we had made the effort to be present. The whole of our entourage had attended on this occasion, except the children, who were left with the Nanny. The adults were able to get into a single car, so Fritz drove us back to the hotel from where we found a small restaurant for a late evening meal.

The Nobel Awards

On Thursday was the big event, the actual award of the Nobel Prizes, when we would all meet in the Royal Theatre, which had been suitably decked out for the occasion. The ceremony was due to start at about four o'clock in the afternoon and was expected to last for a couple of hours. We were bussed over to the Royal Theatre, and the Laureates were taken back stage, where we prepared ourselves for the ceremony. It was at this point that we were to come into contact with the Royal family, although we did not have sight of any of them until we were actually on stage. The Royal family entered from the front of the stage to take their positions.

My wife and our family members were ushered to their reserved seats fairly close to the main stage. With the Royal family seated on stage, the various Laureates were called in turn to receive their Nobel Prize. As he awarded each prize, the King made a few comments of congratulation, and the Nobel Prize medallion was handed to the recipient together with the Nobel Prize certificate. The recipient then returned to his seat, and the next person was called, until all six or seven of the Laureates had been awarded their medals and certificates. This whole process took about half an hour and shortly after the ceremony was over the Royal family left the stage. At that point members of the audience, mainly members of families of the Laureates, came up on stage and congratulated their relatives. So my wife came up on stage to congratulate me, together with my two daughters and their children (see Figure 35). This whole process took maybe another half hour and then at about six o'clock in the evening we were ushered out into coaches and transported to the

Figure 35 On the stage with the family, after the Nobel ceremony, 2003. From left to right: Samantha Mansfield-Murphy, Charlotte Mansfield-Murphy, Thomas Haywood, Matthew Haywood, Sarah, Peter, Jean, Gillian, and Brett.

Reception Hall, which I believe was part of the Town Hall, where the main reception and dinner were held.

As had been agreed beforehand, one Laureate from each of the groups of prize winners was asked to give a short speech lasting no more than five minutes. As Paul Lauterbur was not too well and did not feel up to it, he suggested that I give the speech representing the two of us. So at an appropriate point during the dinner I was called upon to go up to the podium and say my piece.

The dinner itself was a grand affair, and there must have been 150 people sitting down to the meal. My wife found herself sitting next to Al Gore, the American ex-Vice President, and I found myself sitting between two members of the royal family, the young Princess Madeleine on one side and Princess Lilian, Duchess of Halland, on the other side. We were able to talk and quite freely discuss a whole range of topics while we enjoyed the sumptuous meal that was offered.

On the Friday we took up the offers of prearranged visits to various places in Sweden. We went to the University of Upsala, which is some thirty or forty miles outside of Stockholm. One of the people we met in Upsala was Rolf Shurblom, who had previously visited the University of Nottingham for a year while on sabbatical leave. We had also seen Rolf earlier at the hotel in Stockholm, but on our visit to Upsala we were shown around various departments within the university. The President laid on a splendid luncheon for us. We departed in the middle of the afternoon in order to get back to Stockholm in time for a final major event that evening. This was a visit to the Palace itself in Stockholm. This turned out to be another grand dinner, with at least twice as many guests, and it really was another fantastic occasion. On this occasion only the Nobel Laureates and their spouses were invited, but there were many dozens of dignitaries, mainly from Sweden and Norway but also including ambassadors,

Figure 36 Posing with the Swedish royal family, 2003. From left to right: Sarah, Peter, Queen Silvia, King Carl Gustaf, Jean, and Gillian.

consular attachés, and other representatives from other countries, present at the dinner.

At one point in the evening we were led upstairs, and in a rather grand hall within the Palace we were greeted by the Royal family. We met King Carl Gustaf and Queen Silvia personally and had our photographs taken with them (see Figure 36), and it was then that I had occasion to say to the Queen that I had heard that she came from Heidelberg in Germany. Graciously she said a few words to me in German.

Our two daughters, their husbands, and their children had left to return to England on the Friday morning; again Gillian, Brett, the two boys and their Nanny flew back, while Sarah, the two girls, and her husband returned by train. We took our return flight to London on the Saturday morning.

14

Antagonisms to MRI

David Hoult

I first met David Hoult at a British Radio Spectroscopy Group (BRSG) meeting held in Canterbury. I was invited to give a paper on multiple pulse techniques that we were developing at the time in Nottingham, and David, I believe, was a postdoc working in Sir Rex Richard's group at Oxford. David got heavily involved in the technical development of high resolution spectroscopy and later became involved in George Radda's group in the development of spectroscopy and spectroscopic imaging *in vivo*.

In the early 1970s my interests had turned from multiple pulse techniques. I was not especially interested in spectroscopy and concentrated all my efforts on NMR imaging at the time. One of the problems that concerned me very much in the early development of MRI was the question of defining a slice of material. We had decided that this should be done by one or other of the slice selection techniques that Peter Grannell, Allen Garroway and I had developed and patented.

Early in 1978 a pre-print arrived on my desk of a paper by David Hoult entitled 'Zeugmatography: A Criticism of the Concept of a Selective Pulse in the Presence of a Field Gradient'. The paper, which was highly critical of our slice selection approach in MRI, had been sent to the *Journal of Magnetic Resonance*. I immediately corresponded with the editor, Wallace S. Brey, Jr, who informed me that the paper had been accepted for publication.

Although we had tried the technique of selective excitation experimentally and it seemed to work quite well, there was considerable scepticism in the MRI community, and so we decided to write a reply refuting Hoult's paper but also indicating explicit details explaining how our slice selection technique worked. Specific contributions to this reply were made by Andrew Maudsley, Peter Morris, and Ian Pykett. We had by then been regularly using slice selection to produce images of all kinds. The paper by David Hoult appeared in the *Journal of Magnetic Resonance* (26 (1977), 165), stating that we had described a technique that could not possibly work, since it violated a basic principle of physics, namely the Uncertainty Principle. This assertion was clearly nonsense, since we had used slice selection for many months and had demonstrated experimentally that it was a perfectly useful technique. I felt that Hoult's paper was something of an affront, and our somewhat longer reply was a vigorous refutation of his paper. This was published as a full paper in the *Journal of Magnetic Resonance* (33 (1979), 162–74) with the title 'Selective Pulses in NMR Imaging: A Reply to Criticism'.

In the paper, calculations are presented that show in detail the effect of applying selective RF pulses to an inhomogeneously broadened spin system. Experimental results confirming the predictions are also presented, which demonstrate the conceptual correctness of NMR imaging methods based on selective irradiation.

In a letter to me Paul Lauterbur was also critical of Hoult's paper. But two of Paul's students within his research group were not happy with our approach to defining slice selection and questioned what we had written.

Paul Lauterbur

Shortly after my contretemps with David Hoult, Peter Allen decided to take a sabbatical leave of approximately one year to work with Paul Lauterbur in his laboratory at Stoneybrook. This was before Lauterbur had been offered the position in Urbana, Illinois. So far as Paul Lauterbur's group was concerned, the question of slice selection and selective irradiation was still an open topic, and some of his research students were adamant that slice selection as we had described it would not work. On his arrival in the States, Peter Allen spent some of his time giving tutorial talks to the group in Stoneybrook. He demonstrated mathematically that slice selection does work. Peter wrote to me while he was in the States and told me about his problems and the slight difficulties that he was under and rehearsed his arguments to me in a letter. These were correct.

This little episode seemed to me to be symptomatic of the growing rivalry between Paul Lauterbur and me. For example I learnt later that he had been recruited to act on behalf of a company in the States that had tried to convince the patenting authorities in America that our ideas on imaging were incorrect and had tried their hardest to get one of our patents modified or withdrawn altogether. One of the surprising things is that, while we were seeing Paul Lauterbur quite regularly in the early days, he was working with other interested parties in America trying to undermine our patent position. It turned out that Paul Lauterbur had not been as open as he seemed when he came over regularly to visit us at Nottingham. It appears that he was working in a clandestine manner with an industrial concern at the time in an effort to negate what we had already achieved and covered in patents. But it later transpired that the real reason he visited Nottingham so frequently was as a cover for his visits to see Joan Dawson in London. He later divorced his wife to marry Joan.

When I met Paul and his wife Joan in Stockholm at the Palace Hotel, I decided it would be inappropriate to mention the issue concerning patents. I include these details here simply to give an accurate representation of the story of MRI.

Raymond Damadian

The first time I met Raymond Damadian was in 1976 at the Ampère Conference meeting held in Heidelberg. I gave a talk on MRI that lasted about half an hour and in it I spoke initially about the work we had done on imaging fingers and also included other images that we had produced. Raymond Damadian got up to speak after me and spoke for about half an hour, but he did not really have anything comparable in quality to what we had produced. After the talk he was visibly shaken. I remember him walking away from the meeting, which finished at about twelve o'clock, with his entourage of students. I managed to speak to one or two of his students a little later, and they said that he was shocked by what he had heard during my talk and went away depressed and looking most downcast and dejected.

In 1979 I was promoted to a Professorship in Physics. Shortly after that I was invited to Cripps Hall of Residence on the Nottingham University campus. The President of the Hall invited me to attend a dinner, which he explained was a customary event for the Hall to enable new professors to meet the students of the Hall. The whole occasion occurred at the beginning of the Autumn Term of 1979. So I went along to the dinner, and was entertained by students and other guests at the High Table during the course of the evening. At about nine o'clock, when we had reached the point where we were about to have dessert, a porter suddenly came into the crowded hall, approached the top table, and asked for Professor Mansfield. I responded and asked what the problem was. The porter said that there was a fellow outside who wished to speak to me. I was taken aback by this and extremely puzzled to know exactly what had happened and who could be outside wanting to speak to me. So I excused myself from the table and followed the porter into the foyer, and there was Raymond Damadian. I had a quick

Antagonisms to MRI

word with him and asked him what the problem was. He said that he had come to Nottingham especially to see me. He was apologetic for disturbing me at this time in the evening, but it was so important that matters could not wait. I explained to him that I was a guest at the dinner and that I had to return to my hosts, but that in ten or fifteen minutes I would try to make my excuses from the High Table and join him. So I went back into the hall, had the dessert, and then spoke to the President, made my excuses, and slipped out.

When I got outside, I suggested that perhaps the most sensible thing to do would be for Raymond Damadian to accompany me back to my home. At the time we were living in Clarke's Lane in Chilwell. So he followed me in his car and when we got home we offered him some refreshments and sat down to hear what he had to say.

By now the time was 9.30 in the evening, and we started to find out slowly what the problem was. It seemed that Raymond Damadian had come to Nottingham to talk to me specifically about the problems that he was having in the States and how he felt he had been mistreated by a number of institutions there and especially by Paul Lauterbur. The problem, he said, was that he had requested money to do research from the National Science Foundation and also from other institutions, but at every turn his efforts had been thwarted and his research grant applications rejected. He also related to me his version of the story of what had been going on at NMR Specialties, the company in Pittsburgh, and his interaction there. He said that he had been retained as a consultant to NMR Specialties and had made occasional visits to the company to carry out some work on the measurement of spin lattice relaxation times in various types of normal and tumorous tissue. But during the course of his work there the company had fallen into financial difficulties, and Paul Lauterbur had been asked to take over the Chairmanship of the company while Raymond Damadian was still acting as a consultant. He also said that the work that he was doing was well known to various members of the NMR Specialties Group and especially to the new Chairman of the company, Paul Lauterbur.

He then went on to tell me that during the course of his work on the measurement of relaxation times he had had the idea to do imaging by a form of point scanning of the object with a view to looking at the tumorous material in a non-invasive manner. His experiments at the time were, of course, highly invasive, and it was necessary to take samples from the infected animals. Quite often it was necessary to sacrifice the animals, this being the more humane way of taking the necessary samples for NMR investigation. Because of the trauma caused to the animal, he had thought of a way of taking measurements from the living animal by doing a localized point or regional scan experiment. He said that he had discussed this openly with members of the various groups in NMR Specialties, and his ideas were well known and discussed by all people, including Paul Lauterbur. He explained that he had come to England specifically to discuss the matter, and the lack of support from the American funding agencies, with me and to relate his troubles, because he thought that he would receive an understanding and sympathetic hearing of his problems.

I explained to him that I was really in no strong position to help him one way or another but I was very sorry that he had been treated in this way. But of course I had no evidence and no way of knowing whether his version of events was correct or not, so I accepted on face value what he had told me. By then it was getting late. Jean had gone to bed at about 11.30 in the evening, and we continued talking about the whole business and batting around ideas until something like 12.30 or one o'clock in the morning.

At the time we lived in a small house in Chilwell. We had two children and we had no spare room to offer to put him up overnight. However, I did suggest that we could make up a makeshift bed on the sofa if he wished to stay over, but he was quite adamant that he wanted to leave and had to get back to London that night. So at about one o'clock he bade us farewell, got in his car, and drove away.

The third time that I met Raymond Damadian was during a Royal Society meeting that Raymond Andrew had organized in London. I was involved peripherally in the organization and, despite the comments that we had heard about Damadian and his antics with Paul Lauterbur, I argued strongly that he should be invited to come to the Royal Society and present his views. So this was done. He came to the Royal Society meeting and had a half-hour slot to speak. When his time came to speak, he started his lecture by showing a slide of President Reagan and went into a tirade about how he had been persecuted and turned down by the various funding agencies in America. He went on and on complaining about the whole situation and the sniping that was going on in the States. In the end his time ran out and he had hardly spoken at all about the topic he was supposed to address. The Chairman of the session asked Ian Young to go on stage and usher him off while he was still ranting. I was very shocked at this behaviour and really felt that he had not done himself any good whatsoever. If he had an interesting scientific story to tell us and some interesting results, that would have saved the day and restored people's faith in him as a serious scientist. But the behaviour that he exhibited on that day at such a prestigious institution really made everyone, including me, feel that his was a lost cause.

One of Raymond Damadian's principal opponents in those early days was Dr Donald Hollis, who, it turned out, had been one of the grant reviewers on the Biophysical Chemistry Study Section of the National Institutes of Health (NIH). He seemed to know more than most people about Raymond Damadian and his various attempts to achieve government funding. In those early days Hollis attended meetings and gave papers on the work that he was involved in, which was the study of different types of tissues, measurements of $T1$'s and $T2$'s and so on. But at some later stage he decided to retire to take up the running of a private hotel, which I imagine had been left to him. It is possible that he inherited this from his own parents. In his retirement he wrote a booklet that he published. A copy of this book was later sent to me by Professor Burton Muller, then Professor of Physics at the University of Wyoming at Laramie. The book is entitled *Abusing Cancer Science: The Truth about NMR and Cancer* by Donald P. Hollis, published at the Strawberry Fields Press, Chehalis, Washington, in 1987.

The book gives a blow-by-blow account of Hollis's interaction with Damadian, detailed in fifteen chapters with headings such as 'Trashing the Cancer Establishment', 'Conning the Congress', and 'The First Human Image'.

Although one or two other people get a mention in the book, including me, the book is heavily biased against Damadian, and Hollis takes every opportunity to bolster Lauterbur's position as sole inventor and originator of MRI, to the point that his account becomes cloying and wearisome.

To my way of thinking, a broader and more balanced account of the early days in MRI would have assured a place in history for Hollis's book as a historical document.

15

Beyond the Nobel

General Magnetic

Following my retirement, I was able to continue my research activities both at the university, and also with the help of the small company I had founded, General Magnetic, where my son-in-law Brett Haywood was working for me by this time. Brett had originally worked for Central Electricity and obtained a Higher National Certificate (HNC) in Electrical Engineering on day release. I had suggested he might further his studies, and Brett first gained an honours degree from Derby University, and then took a four-year part-time course at the University of Nottingham to pursue a Ph.D. in MRI. The project for his Ph.D. was the continuation of work connected with producing a low noise gradient coil for MRI. This work was finished and his Ph.D. presented in 2006. He managed to produce a gradient set for a model gradient coil system with an object size of 5 cm diameter. This was successful in producing an image and at the same time reducing the acoustic noise level somewhat. However, after his Ph.D., the work continued, and the following year, 2007, a much better image was produced and a much greater noise reduction achieved. This work was published, with Brett Haywood named as the first author with Barry Chapman and me as co-authors. The paper was published in the journal *MAGMA* in 2007.

General Magnetic continued to put a lot of effort into acoustics, especially the problem of noise reduction in magnetic gradient coil design. To this end we designed and constructed a new gradient coil system with the help of Barry Chapman. This was large enough to take a human head. Because of the many delays that we experienced using the old electronics associated with the 3T system, I decided that the sensible approach was to replace it with something new built commercially.

To this end we purchased some commercial imaging equipment from MR Solutions Ltd and a new set of gradient amplifiers from the American company Copley Amplifiers Inc. This company had just been acquired by another American company, Analogic. The new commercial system was ordered and arrived in March 2008. It took some time for the new equipment to be installed and accepted for continued experimental use. In the meantime Barry Chapman's new coil system was made and underwent tests. The hope was to try imaging a volunteer. The aim here was to produce a quiet coil system capable of imaging the brain. We were looking to produce 128^2 pixel EPI snapshot images.

One of the advantages of the new MRI system was that it was very flexible. It appeared at the time that it would be possible not only to use it for imaging but also to try some pulse experiments reverting back to the work on solid echoes. The idea was the possibility of revisiting some of those questions that are still outstanding concerning the production of solid echoes following a 90–τ–180° pulse. There were unexpected effects that I had noticed all those years ago, and this topic has never really been fully explored. The hope was that with the new system we would be able to mount some solid echo experiments and resolve the issues that were raised at the time. One of the problems at the time was that the pulse spectrometer that I had built and used in the early 1960s was incoherent—that is to say, the RF phase of the initial 90° pulse and the second 90° or 180° pulses were unrelated phase-wise, and this created a problem, which I was able to rectify in those early days by very careful timing of the two pulses. However, the new equipment used coherent RF, so that all the pulses have a fixed phase relationship. The phase of a pulse train with respect to the initial 90° pulse could be changed, thereby making it possible to re-examine in much more detail, with a modern RF system, the production of a solid echo following a 90°/180° pulse.

The work that Brett started—namely, the effect of electric fields on patients being imaged—initially indicated that there is a sizeable electric field associated with time-dependent magnetic field gradients. We proposed a way of reducing this electric field. This work was published recently in the journal *Physics in Medicine and Biology*. However, recent theoretical and experimental work by Barry and Brett suggests that the paper in *Physics in Medicine and Biology* is wrong. I hope therefore to return to the e-field question.

The BRSG Meeting in Oxford

My colleague Professor Tony Horsewill of the Physics Department at the University of Nottingham was secretary of the BRSG in 2004 and gave me a lift to the annual meeting that year, held in Oxford.

A number of friends and former colleagues also attended this meeting, which lasted for one and a half days (see Figure 37). Both Jack Powles and John Strange gave papers on aspects of their work in NMR. I gave a short paper on my early work on solid echoes, drawing attention to the unexplained behaviour of these echoes when formed from a $90°_0$—τ—$180°_{0,180}$ pulse sequence.

Flying

Ever since my days at the RPD, I had had a latent desire to fly. So shortly before I retired in 1994 I decided to take it up. I was encouraged to do this because we had a member

Figure 37 John Strange, Ed Randall, Jack Powles, and Peter Mansfield attending the BRSG Conference in Oxford, 2006.

of staff in the Physics Department, Barry Hill, who flew fairly regularly at the Sherwood Flying Club at Tollerton Airport, just close to Nottingham. He had taken me up once or twice for flights around the Nottingham area, and this had reawakened in me the interest in flying.

After my retirement I decided to take the whole thing more seriously and join the flying club. I spent a year learning to fly. Eventually, I obtained my fixed-wing private pilot's licence (PPL) and as soon as I had qualified for that I decided to take up helicopter flying. This meant joining the East Midlands Helicopter Company, where I flew mainly with Nigel Burton, who owns the company. We flew from his farm, based at Costock near Nottingham. It took a little over a year to qualify. My total time learning to fly a helicopter was about sixty hours, whereas I managed to qualify for fixed-wing flying in the recommended forty hours. For fixed-wing flying I flew a lot of the time with Steve Smoothy, one of the instructor pilots at Sherwood Flying Club. I also flew with Len Stapleton, the senior flying instructor, who had been in the Royal Air Force. When he retired, Peter Clark took over as senior flying instructor; sadly he has recently died. All these chaps were excellent teachers and I was able not only to learn to fly but also to understand the necessary theoretical work that goes into taking the PPL.

While I was learning to fly the helicopter I also continued to fly solo on fixed-wing aircraft. On one occasion when there was a slight tail wind along the runway I came in to land a little bit too fast and could not or did not apply the brakes sufficiently, so that unfortunately I rolled off the end of the runway. This resulted in a minor incident that was reported to the Civil Aviation Authority (CAA), but there were no repercussions. I decided at that point that it would probably be best to concentrate on one aircraft. Thereafter I concentrated on the Robinson R22 helicopter. When I finally qualified in the helicopter, I received the private pilot's licence helicopter (PPLH) and continued flying for about a year on and off until 2000, when my prostate problem was diagnosed. I believe I flew for the last time just following the Christmas period in the year 2001, since when I have not flown at all. Of course the PPL and the PPLH are licences for life, but in order to maintain the ability to fly one has to submit to a regular test, particularly when you have not flown for a while. Because of my medical status following the prostate operation and since then an aortic aneurism, it would be necessary for me to get dispensation from the CAA in order to validate my licences. I would then have to undertake a revision flying course in either or both the helicopter and a fixed-wing aircraft in order to re-establish the validity of both licences. Whether I do this or not depends very much on my state of health. But I am quietly looking forward to the possibility of flying again in the not too distant future.

Recent Honours and Awards

In 2006 I received a letter from Sandy Nairn, Director of the National Portrait Gallery in London. He informed me that I had been chosen to sit for my portrait, which, when completed, was to be hung permanently in the gallery. He also invited me to visit the gallery to discuss ideas as to who might be commissioned to paint the portrait. He sent me some National Portrait Gallery publications that illustrated the work of a number of promising and talented young artists.

I spent the day viewing portraits in the galleries and had a nice lunch with Sandy and his assistant and explained to them that my preference was for a realistic and traditional portrait rather than a modernistic portrait where eyes and ears can be placed anywhere on the canvas, as in one of Dali's portraits. Having made clear my preference, I returned to Nottingham to await word of a suitable artist. This was in September 2006.

In January 2007 Sandy informed me that he had approached a young Scottish artist who was prepared to take on the commission. His name is Stephen Shankland, a winner of the *Times* newspaper Artistic Awards in 2006. Shortly afterwards I was contacted directly by Stephen, who was keen to meet me and to make a start in the planning of the portrait. He drove down from Scotland one weekend and we met on the Monday at about 10.00 a.m. Stephen took a number of photographs of me in my office at the university. He brought with him a large mirror and took some reflected images of me at what I considered to be strange angles. Eventually he departed at around 3.00 p.m. and returned to Scotland.

A month or two later I received word from Stephen than he had drafted out a portrait on hardboard and wanted to return to Nottingham to finish the portrait. There was now some rush to finish the work, so that it could be submitted to the Portrait Gallery for their approval and acceptance.

Stephen came down to Nottingham again, but this time he was accompanied by his wife and their new daughter. Again he spent most of the day touching up his draft, until he was satisfied with the shading and certain facial details. The portrait was taken away to be framed, after which it was submitted to the Portrait Gallery.

A few weeks later I learnt from Sandy that it had been accepted and would be hung in the gallery for a private unveiling on 12 September 2008 (see Figure 38). My wife and I, together with up to eight guests, were invited to the unveiling, followed by lunch. Among the other guests present was the Vice Chancellor of Sheffield University, who was on the National Portrait Gallery selection committee and whom I suspect was responsible for selecting me as the subject for the portrait. The whole day went well and was a delightful occasion that will for ever remain in my memory.

In 2007 I received an invitation from the BBC to be a guest on the programme *Desert Island Discs* hosted by Sue Lawley. Our chauffeur, Malcolm, drove me to the BBC recording studio on Portland Place in London. I had previously been asked to select a number of tunes and music, but one of the pieces I had chosen was a German folk song, 'Vom Barette schwank die Feder', sung by Heino, that the BBC did not have in its archive. Fortunately I had an LP disc of this work which I took down with me. Other pieces that I chose were from the works of the English composers Elgar, Holst, and Walton, the Czechoslovakian composer Smetana,

Figure 38 With artist Stephen Shankland at the private viewing and unveiling at the National Portrait Gallery, 2007.

and the French song 'La Mer' sung by Charles Trennet. All told, the event was a delightful affair that I enjoyed very much.

Also in 2007 I was awarded the *Times* Higher Award in the category of Lifetime Achievement especially for my work in the development and application of MRI. The award, an especially commissioned trophy, was made by the Labour peer Baroness Kennedy at the Grosvenor House Hotel, London. In her speech as one of the judges under the various award categories, she commented: 'Here is a man who left school at 15 and ended up winning a Nobel Prize. What an incredible and inspirational story.'

The Pride of Britain awards were organized by the *Daily Mirror* and the ITV network. I was especially interested to learn this, as I had once had occasion to write to the editor of what, at the time, was called the *Children's Mirror*, which was published each Saturday. I mentioned this to the current *Mirror* sub-editor, Peter Willis, who subsequently sent me a xerox copy of the piece from 16 January 1949 that had prompted my letter to the editor.

The Pride of Britain Award was also that of Lifetime Achievement, and the judging panel included the Prime Minister's wife, Sarah Brown, and Chief Medical Officer, Sir Leam Donaldson. I was informed that previous winners included Sir Alec Jefferies, who pioneered genetic finger printing, and Sir John Sulston for his work on the human genome project.

The whole process started on Sunday, 4 October 2009, with an informal dinner at the City Inn Hotel, Westminster, hosted by Carol Vorderman. This was most convenient, as we were staying at this hotel. We had the opportunity to meet some of the other winners in different categories. Quite a few of these were people with serious medical conditions, some terminal.

The following day, Monday, 5 October, was somewhat relaxed. In the evening we were all taken over to the Grosvenor House Hotel on Park Lane for the major event. I was surprised to be called on to the stage to meet Gordon Brown, who introduced me to the audience. After receiving the award I was allowed to say a few words of thanks and recognition of the many colleagues and students who had worked with me over the years.

On Monday, 30 November 2009, I was awarded the MRC's Millennium Medal by the Chief Executive of the MRC, Sir Leszek Borysiewicz. The award, which is made 'in honour of scientists who have made a profound impact of improving human health', was presented to me at a specially convened meeting of family, friends, and colleagues at the University of Nottingham. The meeting was followed by a dinner organized by the MRC and held on the Nottingham Campus.

I learnt during the course of this meeting that Sir Leszek was in the process of leaving the MRC and taking up the Vice Chancellorship of the University of Cambridge.

On Wednesday, 23 June 2011, I was invited by Sir Leszek Borysiewicz to visit Cambridge University to receive the Honorary Degree of Doctor of Science. I travelled to Cambridge with Brett, and we stayed overnight at Hughes Hall, where as a Fellow I was invited by the President, Sarah Squire. Among the other recipients of Honorary Degrees was Milly Dresselhaus, a past President of the American Physical Society and a Fellow of the National Academy of Sciences. It turns out that she knows Charlie Slichter, who is also a Fellow of the National Academy of Sciences.

16

The Epilogue

Family, Friends, and Colleagues

Apart from my immediate family and, of course, Florence and Cecil Rowland, there have been many people who have exercised great influence on me during the course of my life. Of course, quite a few of these people have already been mentioned to a greater or lesser extent in the body of the text. Some have been mentioned in passing, without amplification, but others who have had an especially strong influence either on me directly or on my career merit particular attention. It is, therefore, my main object in this epilogue to do justice to those friends and relations who have helped me or have influenced me over the years, either personally or professionally.

Conrad William Mansfield

Conrad, my eldest brother, was born in 1924. Con left school in 1938 and worked at a small engineering and fabricating company on Picton Street, diagonally opposite to where we lived. He continued working there until he was 17 years old, and in 1943 volunteered to join the Fleet Air Arm, where he served on an aircraft carrier as an aircraft engine mechanic (see Figure 39). Towards the end of the war his aircraft carrier visited Kowloon in Hong Kong. It was while he was visiting Hong Kong that he met a Chinese girl, Isobel, half of whose family had emigrated to the United States before the

Figure 39 My eldest brother, Conrad William Mansfield, aged 18 during the Second World War.

Japanese had invaded. After the war she herself decided to go to the States.

Shortly after the war's end, Con's ship came back to England and docked at the Chatham port, and it was while he was stationed there that he met his wife, Lena. When he was finally demobbed, he set up home with Lena, and they lived in Morden, south London. Within a year or so they had their first child, Carolyn, but Con was not satisfied with the way that things were working out in Britain shortly after the war, so he and Lena decided that they would emigrate to Canada. He explained to us all that he would go to Canada first to find a place to live and to find some work; then he would arrange for Lena to follow him within six months or so. So he emigrated, but within a few months, we found out subsequently, he had tried to cross the border from Canada into the United States and was picked up by the immigration authorities in America and was sent to Ellis Island.

We found out later that, before being apprehended, he had made contact with Isobel, who had left Hong Kong and joined her uncle, who was running a restaurant in New York. After some weeks, Conrad was deported to England, and shortly afterwards we learnt that Isobel had made her way back from New York and joined Con in London. In the meantime, Lena had gone off to follow Con, not knowing exactly where he was or what was going on. She went to where she thought he was, but of course he was not there. He was detained on Ellis Island at the time, awaiting deportation to Britain. They never met up. Lena wrote back to my parents saying that she had decided to seek a divorce. Within a year or so she was divorced from Conrad, and by then Con was living in London with Isobel.

Following the divorce, Lena settled in Canada and wrote the occasional line to my parents. After a while she let them know that she had met someone else in the States and was proposing to remarry, and that really was the last we heard of Lena, or so I believed. However, in December 2006 I received a letter from a certain Helena Paradise, who turned out to be Con's ex-wife Lena. Her daughter Carolyn had been suffering from a brain tumour, so she had been searching the Internet for more medical information and in the process had stumbled upon my connection with MRI. As a result she wrote to me with her news. Sadly Carolyn died in 2007 aged 59.

Conrad, meanwhile, had re-established himself in an engineering company, although we never knew the full details. At one point he claimed to be a partner in a small engineering company, where he was engaged as a tool-maker. In the meantime, Isobel, whose English was improving, decided to set up a hairdressing salon in New Cross, south London.

As a boy Conrad had had a wonderful singing voice. He sang in a local Sunday school choir and he also won a number of competitions for singing. When his voice broke, he became a rather good tenor singer, but he never took this up professionally. However, he was often asked to sing at meetings or gatherings of one sort or another. I remember him telling us once that he and Isobel had taken a slow boat, literally, to China. I believe she was visiting relatives in Hong Kong. While on the boat, Con had entertained some of the guests by singing. One of the songs he sang was 'Retorna a Sorrento', which he apparently sang in Italian. Although there were Italians in the audience, it was said that his singing was so good that they overlooked the fact that the Italian was not completely correct.

Conrad stayed in the engineering and tool-making business until he retired. Towards the end of his life he was diagnosed with cancer in one of his kidneys and that was removed. Later he had a problem with his stomach that necessitated more surgery. This was successful, but one of the aftereffects was that he suffered intestinal adhesion. This necessitated a further operation. This was such a trauma for him after two major operations that at one point he decided that he really did not want to continue living, even though we all tried to reassure him that he should be fine now that his intestines had been fixed. But he lost both his appetite and his will to live and gradually faded away. He decided that he did not want any more dealing with medicos or hospitals, and died in 1992 at the age of 68.

He had two children by Isobel, Philip and Jacklyn. His funeral was held in Carshalton, and he is buried in the cemetery there. Isobel died in 1996 aged 74, and she is also buried in Carshalton.

Sidney Albert Mansfield

My brother Sid was born on 3 January 1927. It was Sid who spent some years with me in Devon. He left Torquay and the Rowlands in 1943 and returned to London, where he worked for E. R. Watts & Sons until 1944. It was then that he volunteered to join the Royal Navy, sometime around February 1944.

He went to the recruiting office in Croydon with a friend, Arthur Smith. When he was being interviewed, the recruiting officer asked Sid what term he would like to serve in the Navy. Sid said twenty-one years, but the officer advised him to join for the hostilities only. He then went to Chatham, his home base. Following Chatham, he was sent to various land bases. His initial training was in Malvern in Worcestershire.

He had been in the Dulwich Sea Cadets for a few years before joining the Navy, so he was able to tie knots, do drills, and show other skills associated with the Navy. He used to take over drilling the squad from time to time, and the people in Malvern wanted him to stay as an instructor. But the higher ranks disagreed. So after his initial training, he went back to Chatham. When he first joined up, he wanted to be a signaler, but because the signals branch was full, he was sent as a stoker in the engine room. But, of course, by then a stoker did not shovel coal: the ships were powered by diesel engines.

For stoker training, he was sent to Devonport in Plymouth on a training ship, HMS *Imperuse*, and then back to Chatham to await a troop ship: a converted refrigerated food carrier called the *Dominion Monarch*. He had to go to Liverpool to pick up the ship together with a number of other troops, from all three services. The ship sailed from Liverpool, through the Panama Canal, across the Pacific Ocean to Sydney, Australia. While in Sydney, he was transferred to the aircraft carrier HMS *Indefatigable*. On the *Indefatigable*, he saw some action from Japanese suicide dive-bombers or *kamikaze* (divine wind). One hit the deck by the conning tower, and did a little damage, but there were steel decks on the *Indefatigable*. The American aircraft carriers had wooden decks, so they would have been more susceptible to damage.

At the end of the war, his ship sailed into Tokyo Harbour. They had been at sea for three months, and one day the Fleet Commander came aboard and said that a special bomb had been dropped in Japan, and the war was over. Nobody knew what the special bomb was. After a few days they sailed into Tokyo Harbour to witness the Japanese surrender, which took place on the American battleship *Missouri*. Sid also went ashore to see Hiroshima. He said that it was a terrible sight. Complete desolation!

Finally his ship headed back to England via the Suez Canal, stopping off in Ceylon. He brought two large wooden boxes containing tea, for which Mum was most grateful, since food was still rationed in Britain.

Sid was demobbed in 1947 and went back to E. R. Watts. Sid now lives in Marldon just outside Paignton in Devon, and was 85 in January 2012. He lives with his second wife, Gisela, a German national who was originally married to a British serviceman and came to Britain shortly after the end of the Second World War. They met at Sid's firm, Ranks.

After the war Sid tried setting up his own business with a friend John. The business, brush manufacturers, was called Sidjon's and was based in an old disused stable of the former Eagle Mews yard on Elmington Road, in Camberwell. They were able to rebristle old brushes as well as make new brushes for industrial use.

Professor Erwin Hahn

My personal interaction with Erwin Hahn occurred in 1990, when I persuaded him to join me in the organization of a conference held at the Royal Society. It was actually a Royal Society discussion group meeting entitled 'NMR Imaging'.

In point of fact, Erwin is one of the true founder fathers of NMR. I came across his name very early on in my own studies when I was a student. Erwin had published many important papers on the basic principles of NMR. Later on, when I was a postdoc in the University of Illinois at Urbana, I was looking in the internal library of the Physics Department one day and spotted his Ph.D. thesis on the shelf. It indicated that he had in fact done his research and Ph.D. studies in the Physics Department at Urbana.

At a very early stage in the development of NMR he made important discoveries with the so-called spin echo, which is the basis of very many experiments that followed. In particular, the spin echo forms the foundation of much of what goes on in MRI today. I was especially delighted when in the year 2000 Erwin Hahn was elected a Fellow of the Royal Society—an election in which, I am proud to say, I played some small part.

In the Royal Society book *NMR Imaging*, which Erwin and I organized and edited, he contributed a paper entitled 'NMR and MRI in Retrospect'. This paper is the first in the book, and characteristically Erwin plays down his part in the process of the evolution and development of MRI. But I can say categorically that, without Erwin Hahn's contribution to the principles of spin echoes, there would be no MRI today.

As I have already said at some length, Raymond Damadian nurses a grievance that he was left out of the Nobel Prize, but in my view the person who really missed out was Erwin Hahn, since his contributions were and remain the cornerstone to the whole concept and implementation of MRI as it is used, not only in ultra-high speed imaging of the type with which I have been personally connected but also with the many general aspects of MRI as they have evolved and as they currently exist today.

This, in my view, is the true tragedy, despite the fact that Erwin Hahn himself says in his chapter in *NMR Imaging*:

> I take this opportunity to apologise to MRI pioneers in the audience because I never believed that MRI would work... However, I was only one of many unbelievers. Another infidel in particular was Anatol Abragam [who died in 2012], a distinguished French Physics researcher in magnetic resonance. He notes in his autobiography that French clinicians began to buy his book entitled *Principles of Magnetic Resonance*, thinking it would enlighten them in their speciality of MRI. The French Society of Radiology wanted to award Abragam a medal in spite of the fact that he told them he had not contributed to MRI and did not believe it would work.

This story reminds me of an encounter that I had with Abragam in the late 1960s. I was at a conference, I think it was an Ampère meeting, at one of the big cities in Europe, and I bumped into him in a corridor on the way to one of the talks. We exchanged a few words, and I recall asking him at the time whether he was planning to revise his book, which had then been out for some ten years or so. His answer was, in essence, that he felt everything in NMR was covered in his book already, and it did not require any revision. This was said despite the fact that at that very conference there had been a whole range of topics, including my own work on multiple-pulse solid echoes and pulse trains for line narrowing in solids, that he completely disregarded. He considered it all to have been covered already.

Professor Brian Worthington

When I first met Brian Worthington he told me, in his own words, that he was in fact 'the naked reader'. What that meant was that he had no department of his own but was

The Epilogue

attached to the Department of Human Morphology over in the medical school. He was in fact a neuro-radiologist, but there was no department of neuro-radiology in the medical school at the time and he was therefore effectively seconded to Professor Coupland's department.

I first met Professor Rex Coupland in the tea room of the Physics Department. He had been invited by Raymond Andrew to come over to the Physics Department and was taking coffee with Raymond. I came along for coffee and joined them at the table. Rex mentioned that he had recently acquired a new reader in the Department of Human Morphology, a Dr Brian Worthington. As I got more and more involved with Rex Coupland, I had occasion, fairly regularly, to meet Brian Worthington, although at the time his interests were more connected with the imaging work that was going on in Raymond Andrew's group. For several years his efforts were really directed towards that work, and he showed at the beginning relatively little interest in the work that we were doing on imaging.

One might say that he kept a watching brief on our developments and as time went on he got more and more interested in what we were doing. This coincided pretty well with the demise of the Andrew and Moore groups.

Brian's interest in our group started in earnest when Raymond Andrew and Bill Moore left the department. It was then that Brian expressed an interest in collaborating with us. As a result of this collaboration, we published quite a few papers on imaging together and attended the same international MRI conferences, where Brian was able to present some of the clinical aspects of our work. As our interaction grew, Brian became more and more convinced that EPI was the way to go, and he spent quite a lot of his time travelling the continent and other parts of the world spreading the word. He was in fact our champion and promoted the technique of EPI wherever he went. As time went by he was eventually promoted to Professor of Radiology and had his own small department in the Queens Medical Centre.

I recall on one occasion in the early 1990s we had both been invited as speakers at a meeting by the European Society of Magnetic Resonance in Medicine; I believe the meeting was in the Belgium City of Leuven. As part of the social programe, members of the conference were invited to a walk around the town one evening. We went together to view various parts of the city, including the cathedral. While we were admiring the exterior of the building, Brian suddenly spotted Professor Ringertz and said to me quickly 'this fellow is on the Nobel Committee, I think you should meet him'. We stood there, and Ringertz, who knew Brian very well, came over and I was introduced to him. But, because he was a member of the Nobel Committee, I quickly excused myself and continued admiring the architecture of the cathedral. Brian told me later that he knew Professor Ringertz, because he, like Brian, was a neuro-radiologist. Brian would make fairly regular visits to the MR Centre and as a result I decided that it would be sensible to designate a room for both Brian and Rex Coupland. This we did, and his name and Rex's were displayed on the door. That was their room, and they could come over whenever they wished to discuss matters or to write papers or whatever they wanted to do. This is what we did, and Brian availed himself of this arrangement quite regularly.

Brian retired aged 60 in 1999, but continued to visit me on a regular basis, giving me updates on various aspects on our work. He had a great interest in the Icelandic language and could speak it quite well but decided following retirement that he would take a special course and get a degree in the topic. So he enrolled at one of the London colleges and took a degree, I believe part-time. This meant that we did not see him regularly for a while, though he continued to drop in from time to time. I remember Brian telling me that the

Icelandic Prime Minister had been invited to Nottingham and he was a member of the party that received her. When he was introduced to her, he spoke a few words in Icelandic. Apparently she was extremely impressed by this and commented on his command of the language.

One of the crowning moments for Brian was when he was elected Fellow of the Royal Society. I had proposed him as a candidate for Fellowship a year or so before, and I was especially pleased that his case had received such relatively early consideration. Brian came to see me after his ceremony at the Royal Society and mentioned that he would very much like to be awarded the Gold Medal of the Royal College of Radiology (RCR). If awarded, this would be the pinnacle of his career in radiology. I said that I would do whatever I could to help him achieve this, especially since he was now a Fellow of the Royal Society. I wrote to the President of the RCR, who suggested that I send him a testimonial and the necessary background information and *curriculum vitae*. I was told that the Medals and Awards Committee would consider Brian's case for receipt of the Gold Medal at their next meeting, and I was delighted when I was informed that the RCR had decided to give Brian the award. As an Honorary Fellow, I was invited to give the oration at the Society's premises in London.

Professor Donald Longmore

Professor Longmore had heard me speak at a conference in one of the London colleges and decided that he wanted to speak to me privately. Shortly afterwards I received a phone call from Donald to ask if he could come to visit me in Nottingham. I invited him to come up, and we spent a morning looking around the old centre in the Physics Department and discussing MRI. It turned out that he was extremely interested in the ultra-high-speed echo-planar technique that we had been developing and of which I had spoken at the conference. So he very quickly became a champion for ultra-high-speed imaging. He was so impressed with what he saw and heard that shortly afterwards I was invited to a cardiac meeting in York.

Donald Longmore was a consultant clinical physiologist at the National Heart Hospital in London and he organized an International Symposium that was held at the University of York in April 1980. Donald had kindly invited me to give a paper at this meeting; the title of my talk was 'The Perioperative Value of Nuclear Magnetic Resonance'. Although we had produced some EPI data in the early 1980s, it was early days, and the real thrust for EPI came a little bit later than that. Nevertheless Donald kept in touch from time to time and seemed to take a great interest in our particular role in the development of MRI. This is doubtless because of his work in cardiac surgery; he was able, from an expert point of view, to see the advantages of ultra-high-speed imaging. As time went by he became more and more interested in MRI. Through his interaction with Picker International, he obtained an MRI scanner at the National Heart Hospital, but continued to be interested in the EPI technique. At one point he was thinking about setting up his own laboratory to make an EPI imager. He consulted me several times on the design and construction of suitable gradient coils to do this and was even thinking about making a very large scanner comprising the whole room. His idea was to create an arrangement where patients would walk in, sit down, and be imaged in the room itself, the room comprising a huge magnet. This, of course, was an idea that never really matured into anything tangible.

Although Donald had no professional background in physics or engineering to my knowledge, he managed to accumulate around him a small group of physicists who were able to take some of his ideas and implement them to some extent. His major idea, however, never saw the light of day. Nevertheless I greatly value his friendship and admire his continued enthusiasm and staunch support for MRI.

Professor Charles Pence Slichter

I have already mentioned the kindness shown to me and Jean by Charlie and Nini Slichter on our arrival in Urbana in 1962 as well as the scientific supervision and leadership that he offered on scientific matters. Now, however, I want to mention an event that occurred more recently in connection with an invitation I had from the University of Leipzig.

In a letter from Professor Juergen Haase received in 2006, I was invited to accept an honorary degree in Physics. As it turned out, Professor Haase had recently returned to Germany and taken up his position at the university. Prior to his return, he had spent some time in the States working with Charlie Slichter. I was pleasantly surprised to find that Charlie, who was still collaborating with Professor Haase, had brought forward his intended visit to coincide with my honorary degree. As a result, we were driven to Leipzig by our chauffeur, Malcolm Parker. When we arrived in Leipzig we learnt that Charlie was already there with his second wife, Anne (see Figure 40).

The honorary degree ceremony, conducted in Latin, came as something of a surprise to me. I said nothing during the ceremony, but I did say a few brief words of thanks at the end. After the formal ceremony, there followed a dinner within the university building to which Malcolm was invited.

The following day we spent some time with the Slichters and later on we had a look around the city of Leipzig. It was full of new buildings and bright and lively squares, quite different from the old and depressing city that we remembered from our earlier visit, back in 1972. Then we had been guests of Professor Harry Pfeiffer. This time, before the degree

Figure 40 Professor Charles Slichter in Leipzig, 2006.

ceremony I had received a telephone call from Harry apologizing for the fact that he was unable to attend the ceremony because his wife was ill. But all told, the visit was most interesting and worthwhile, not only for the honorary degree that had been bestowed on me but especially for the opportunity for me and Jean to meet up with the Slichters after such a long time.

Professor Jack G. Powles

During the course of my career at Nottingham, I have kept in touch with my old research supervisor quite regularly over the years. From those halcyon days as a research student, when Jean and I were regularly invited to Jack and Jill's summer garden parties in north London, to those gatherings in their farmhouse in Canterbury where we met up with old colleagues and former students, these occasions have remained with us both as pleasant memories.

The Epilogue

Since those early days, several events require special mention. The first was on the occasion of my sixty-fifth birthday, when Peter Morris organized a special one-day conference in my honour (see Figure 41). A number of former colleagues and distinguished guests were invited, including the Nobel Laureate, Richard Ernst, and Jack and Jill Powles. At the celebratory dinner held at Prestwold Hall, there were a number of impromptu speeches made, including one from Jack. This rambled on for a bit, with reminiscences from his career as well from my early days. It was very timely and greatly appreciated by me.

In 2002 Jean and I received an invitation to Jack Powles's eightieth birthday celebrations. The gathering was held at a small hotel near to the Powles's new home in south Wales. I was asked to say a few words about Jack and share a few amusing anecdotes, which I was delighted and honoured to

Figure 41 Peter Morris chatting with me at the special one-day conference to mark my sixty-fifth birthday, 1998.

do. John and Annette Strange were also present, together with a number of Jack's former colleagues from the University of Kent at Canterbury. After the formal dinner we were invited over to see the Powles's new residence in the town of Burry Port and to meet one or two of his new colleagues from the University of Swansea. All in all it was a wonderful and memorable occasion to share with Jack and Jill.

APPENDIX

Elementary Principles of MRI

There are four main variants of magnetic resonance imaging (MRI): (1) point scanning, (2) line scanning, (3) planar imaging, and (4) volumar imaging. All other MRI methods are variants or combinations of these four basic methods. In addition to these four there is the possibility of performing scans in either real space or reciprocal lattice space, often referred to as r-space or k-space. Many years ago Nobel Laureate Professor Richard Ernst showed that in nuclear magnetic resonance (NMR) it is better, from the signal/noise viewpoint, to work in the time domain rather than the frequency domain. In my own work on imaging, I have always worked in k-space.

The great French mathematician Joseph Fourier (1768–1830) invented a mathematical transformation, nowadays called the Fourier transform, that allows calculations performed in normal space, or r-space, to be transformed to an equivalent calculation in k-space, and vice versa.

In standard NMR, nuclear signals from an assembly of spins in a liquid, or liquid-like sample, may be observed as a function of angular frequency ω, the absorption line shape, or as a function of time t, the transient signal. Again, signals in the two domains ω, t, referred to as conjugate variables, are related by the Fourier transform. That is to say

$$S(\omega) = \int dt\, \rho(\omega)\, e^{i\omega t}. \qquad (1)$$

This is the standard Fourier transform used in NMR spectroscopy. However, in MRI we require a slightly modified

228   Appendix

expression, because now our conjugate variables are position r and wave vector k. That is to say

$$S(k) = \int dr \rho(r) \, e^{ik \cdot r} \qquad (2)$$

where

$$k = \int \gamma G(t') \, dt' \qquad (3)$$

in which $\rho(r)$ is the spin density as a function of position r, $G(t)$ is the time-dependent magnetic field gradient, and γ is the magneto-gyric ratio of the particular spin species under observation. For example, protons have a γ given by

$$\omega = \gamma B \qquad (4)$$

where $\gamma = 2\pi \times 42.577$ MHz/T.

So far as imaging is concerned, it is always efficacious to use planar imaging over either point scanning or line scanning. Indeed, the most efficacious method of all is volumar imaging. However, to employ this technique requires an exceptionally fast computer with large memory in order to deal with the large data arrays involved. For this reason, echo volumar imaging (EVI) has so far been limited to data arrays describing eight contiguous planes, each plane being 1 cm thick and comprising 64×64 voxels. In this arrangement a Fourier transform of 32 k points was required taking approximately 90 msec. For the whole body images obtained using EVI, the typical voxel size was $5 \times 5 \times 10$ mm³. However, an optimal voxel size to aim for would be 8 mm³, i.e. a factor of 32 smaller than currently achieved in 90 msecs. Ideally, the total imaging time should be kept at around 32 msec to circumvent T_1, T_2 effects. These requirements are not practical with currently available equipment.

For example, with a voxel size of 8 mm³ and a total imaging time $t = 32$ msec scaling the currently achievable EVI parameters, we have $t = 32\,\text{msec} = 3.2 \times 10^4\,\mu\,\text{sec} = 1.28 \times 10^5\,\tau\mu\,\text{sec}$, giving the imaging time per voxel element $\tau = 0.25\,\mu\,\text{sec}$.

Definitions

T_1	is the spin–lattice relaxation time. For human tissue at body temperature, this lies typically in the range 100 sec–1 sec.
T_2	is the spin–spin interaction time. In biological tissue at normal body temperature, $T_2 \leq T_1$.
$\rho(\omega), \rho(r)$	are tissue densities lying typically in the range 0.8–1.0 gm cm^{-3}.
t, t'	are times
ω	is angular frequency
i	$= \sqrt{-1}$
$S(\omega), S(k)$	are received nuclear signals
k	is the wave vector
B	is the magnetic field strength
r	is the position vector
γ	is the magneto gyric ratio $= 2\pi f$
f	is the spin resonance frequency in a magnetic field of 1 Tesla. For protons this corresponds to a frequency $f = 42.577$ MHz.
G	is a magnetic gradient vector
e	$= 2.8$
MHz	is mega Hertz
T	is Tesla $= 10^4$ Gauss $=$ magnetic field strength

Index

Note: *italic* indicates photographs; *fn* indicates footnotes.

Abragam, Anatol 218
acoustic research 172–3, 175–7
active magnetic screening
  148–51, 158
Adams, Mr 28
Advance Nuclear Magnetic
  Resonance
  (ANMR) 142, 143
Ailion, David 67
Alabama University 145–6
Allen, Peter *121*, 134–5, 194–5
Al-Mugheiry,
  Mohammed 173, 174
Ampère meetings 60–1, 93, 97,
  100–1, 109–10
Anderson shelter 14*fn*
Andrew, Raymond 74, 119, 120,
  121, 122, 128, 129–30, 134,
  143–4
Anti-Nobel Prize Meeting 186
Asick, Joe 67
Atala, Ergin *138*
Audley Park School 20

Babbacombe, evacuation
  to 7–17, 20–5
Baines, Terry *113*, *121*, 131–2
Baker, Dick *49*, *50*
Baker, Mr 48
Bangert, Volker *136*
Bardeen, John 68
Barker, Ben *42*
Barker, Colin *168*
Bauer, Ingrid, and Thilo 94–5
Beaumont, Joanna 172, 173
Beeson, Robin *42*

Bellingham, Meg *42*
Bencsik, Martin 173, 174–5
Bicester posting 39–41
Blackband, Stephen 173
Blacksburg conference 123, 126
Blain, Barry *49*
Blay, Alan 115, 116
Blicharski, Professor 97
Blunt, Graham 155, 156, 158
Borough Polytechnic 33–4
Borysiewicz, Leszek 209, 210
Bottomley, Paul 131, *138*
Bottomley, Virginia 166
Bowley, Roger 150, 178
Bowman-Grey School of
  Medicine
  conference 137, *138*
Bowtell, Richard 172, 173
Boys Brigade 13–14
Bradbury, Philip *49*
brain imaging 152, 202
Brey, Wallace S. Jr 193
British Association 165
British Petroleum Company 173
British Radio Spectroscopy
  Group 203, *204*
British Technology Group
  162–3, 178, 179
Brock, Georg *42*
Bronshill Road School 11
Brown, Gordon 163, 209
Brown, Sarah 209
Brunner, Hermann 91, 92, 95
Buchler, Lt *42*
Budapest visit 97
Budinger, Tom 125, *138*

232   Index

Bulford Camp, Somerset 38–9
Buonanno, Ferdi *138*
Burton, Nigel 205
Butler, Frank 36
Butterfield, Professor 128
Buzz Bomb (V-1) 17, 18–19
Bydder, Graeme *138*

Cailly family 30–1, 56, *56*
Cairnes, Mr 3
calcium fluoride
  experiments 79–80
Cambridge University, honorary
  doctorate 210
Campbell, Colin 165–6, 167–8,
  168, 178–9
cardiac imaging 137
Carl Gustaf, King of
  Sweden 192, *192*
Carley-Pocock, Captain 39, 40
Cawley, Martin 149, 150
Challis, Laurie 99, 117–18
Chapman, Barry 148, 149, 176,
  178, 201, 202, 203
chemical shift imaging 152–3
Children's Country Holidays
  Fund 3
Choumert Road School 28
Chrispin, Alan 137
Clark, Peter 205
Clarke, Lenny 34
Clarke, Tony *49*
Clow, Hugh 115, *121*, 129
Cockington Village 15–16
Coleman, Iden 57, 60
comic (The Whizzer) 29
commercial systems 141–2, 143
Conroy, David 47
Cooper, Leon Neil 68
cordite propellants 43–4
Cork Street School 27
Cormack, Allan 118

Coumins, Jaques 178
Coupland, Rex 119, 120, 184,
  219, 220
Coxon, Ron 132, 152
Creyghton, Joris *121*
Cripps Hall dinner 196–7
Crookes, Dr 35, 42, *42*
Crooks, Larry *138*
crystallography
  experiments 102–7
Cuban missile crisis 64–5
Cullis, Roger *49*
Cutler, Doug 57, 58, 61, 66, 70
Cutler, Pat 66

Damadian, Raymond *138*, 185,
  186, 196–200
Dartmouth visit 24
Davis, Derek 158, 159
Davis, Norman 155, 156, 158
Davis, Owen 48
Davis, R. B. *42*
Dawson, Joan 195
Day, John *49*
Department of Health 147, 148
Derbyshire, Bill 99, 111, *121*,
  134, 135
Desert Island Discs 207–8
Devon, evacuation to 7–17,
  20–5
Dickinson, Lionel 42, *42*, 45
Dobbs, Roland 48
Donaldson, Liam 209
Doney, Miss 35
Doodlebugs (V-1) 17, 18–19
doped copper 68–9
double resonance
  spectrometer 69
Downing Street
  meeting 168, *168*
Dresselhaus, Milly 210
Dudman, Eddy 35–7, 51, 148

Index

Dudman, Pat 148
Dyson expansion 60

early retirement 133–5, 169
echo planar imaging (EPI) 120, 136, 137, 142–3
echo volumar imaging (EVI) 151–2, 228–9
Ede and Fisher apprenticeship 32
Edelstein, Bill *138*, 158
11+ examination 27
EMI Central Research Laboratories 115–117, 129
Erlangen works and lab 132–3
Ernst, Richard 120, 121, *121*, 225, 227
Errington, Dr 35, 42, *42*
evacuation
 to Devon 7–17, 20–5
 to Sevenoaks 4–7
Evans, Frank *49*
evening classes 33–4

Fenn, Ruth *49*
film shows 21–2
finger imaging 113–14, 119
Finney, Alwyn 88
fireworks 30, 36
flying 203–6
Fourier transform 227
Fox, Alan *49*, *50*
Fradin, Dr 71*fn*
Franz, Judy 67
French penfriend 30–1
functional imaging 142, 152

Ganssen, Alex 132
Garroway, Allen 81, 86, 87, 88, 102, 112, *171*, 185
Garroway, Mary 86, *171*

General Magnetic 175, 176, 201–3
Gibson, Athol 127
Gilbert, Norman *49*
Glover, Paul 152, 172
Gold, Mr 28
Golden Years 135, 141
Gore, John *138*
Gowland, Penny 137, 139
gradient coil construction 88
Grannell, Peter 81, 87, 88, 93, 112
 correspondence during Heidelberg sabbatical 91, 101–7
Great Yarmouth holiday 32, *33*
Griffin, Bob 81
Guilfoyle, David 152
Gutowsky, Herb 99
gypsum 60

Haar, Professor 141
Haase, Juergen 223
Häberlen, Ulrich 81, 85, 90, 92
Hahn, Erwin 216–18
Hanley, Peter *138*
Harbord, Jean *42*
Hardacre, Col. *42*, 43
Harrison, Eric *42*
Hartland, Tony 57, *58*, 74
Harvey, Ian *168*
Harvey, Paul 151
Hauser, Karl 85, 94
Haywood, Brett 175–6, 183, 185, *190*, 192, 201, 203, 210
Haywood, Matthew 190
Haywood, Thomas 190
head scans 129, 130
Heasty, Dr 48
Heidelberg sabbatical 85–6, 89–107

Hennel, Professor 97, 99
Hermiston, Sqd Ldr *42*
Heron, Ray *42*
Herring, Mr 28
Hext, Donald 14
Hickson, Gordon 148
Hill, Barry *136*, 204
Hinshaw, Waldo 111, *121*, *138*, 155, 187
Holland, Neil *121*, 130
Hollis, Donald 187, 199–200
Honeywell computer 86, *113*
honorary degrees 210, 223
Horn, David *49*
Horsewill, Tony 203
Hoult, David *138*, 193–4
Hounsfield, Godfrey 115, 116, *117*, 117–18
Howes, Albert 29, 35
Hull, Professor 169, *170*
human imaging 113–14, 119, 122–6, 129, 130, 137, 139, 141, 142, 152, 202
Hutchison, Jim *121*, *138*
Hykin, Jonathan 152

Ignoble Prize Meeting 186
Illinois University, Urbana 61, 64–75, 145–6
International Society for Magnetic Resonance (ISMAR) conference, Bombay 111
Interplanetary Society 48–9
Issa, Bashar 173

Jasinski, Andrej *121*
Jefferies, Alec 209
J K Flip Flops 52
Johnson and Johnson 143, 145
Jonas, Jiri 99
Jones, G. O. 48

Kadenoff, Leon 68
Karpf, Armin 53, *54*
Karstaedt, Nolan *138*
Kennedy, Baroness 208
Kent
  evacuation to Sevenoaks 4–7
  holiday to 3–4, *5*
Kent's Cavern 24–5
Kibble, Jean 45, 52, 53, *53*, *54*, *55*, 63, *63*; *see also* Mansfield, Jean
Kibble, Ma and Pa *63*
Kitson, Avril *49*
knighthood 168–9, *170*, *171*
Krakow, Ampère meeting 93, 97, 100–1, 109–10
k-space 87, 109, 110, 227

language skills 28, 31, 93
Lauterbur, Paul 109–10, 111, *121*, *138*, 183, 187–8, *188*, 191, 194–5
Lawley, Sue 207
lead toy soldiers 29
Leipzig University 96–7, 223
Leipzig visit 96–7
Lethbridge, Miss 27
Ley, Willie 133
Light, Mr 28, 30
line scan imaging 112–13, 114
lino cutting 29
Livemore, Laurie *42*
Lloyd, Charlie 35
Locher, Rob *121*
Longmore, Donald 221–3
Luiten, André *138*
Lurey, Fred 67

McCarton, John 155
McConnell, Jack 145
McVittie, G. C. 74

Index

magic pulse sequence 85
Magnetic Resonance (MR)
 Centre 166–7
magnetic resonance imaging,
 elementary
 principles 227–9
Major, John 168
Mansfield, Carolyn 212, 213
Mansfield, Conrad William
 (Connie) 2, 7, 169,
 211–14, *212*
Mansfield, Gillian Samantha 81,
 91, 94, 100, *171*, 185, *190*,
 192, *192*
Mansfield, Gisela 216
Mansfield, Harry 1
Mansfield, Isobel 211, 213, 214
Mansfield, Jacklyn 214
Mansfield, Jean *63*, 65–6, 72,
 78, 94, *171*, *190*, *192*;
 see also Kibble, Jean
Mansfield, Lena 212, 213
Mansfield, Linda *63*, 98
Mansfield, Margaret 98
Mansfield, Mary 1
Mansfield, Peter
 acoustic research 172–3,
 175–7
 active magnetic screening
 problem 148–51, 158
 apprenticeships in print
 industry 32
 army commission selection
 board 40–1
 arrival and early work at
 Nottingham 74, 75, 77–88
 arrival in Stockholm for Nobel
 award ceremony 185–6
 at Blacksburg conference 126
 at Bowman-Grey School of
 Medicine
 conference 137, *138*

 in Boys Brigade 13–14
 at British Association meeting,
 Oxford 165
 brothers 2
 at BRSG meeting 203, *204*
 Budapest visit 97
 careers interview at
 school 31–2
 CCHF holiday in Kent 3–4, *5*
 children 81
 Chilwell house 77–8
 comic production (The
 Whizzer) 29
 correspondence with Grannell
 during Heidelberg
 sabbatical 91, 101–7
 at Cripps Hall dinner 196–7
 Dartmouth visit 24
 date and place of birth 1
 on *Desert Island*
 Discs 207–8
 at Downing Street *168*
 early retirement 169, *171*
 Eastern European visits 96–8
 echo planar imaging talk 120
 echo volumar imaging 151–2
 at eightieth birthday
 celebration of
 Powles 225–6
 at EMI
 laboratories 115–17, 129
 evacuation to Devon 7–17,
 10, 20–5
 evacuation to Sevenoaks 4–7
 evening classes at Borough
 Polytechnic 33–4
 failing 11+ exam 27
 Fellowship of Royal
 Society 120
 film shows 21–2
 fireworks and explosives
 production 30, 36

Mansfield, Peter (*cont.*)
  first MRI meeting,
    Nottingham 120, *121*
  flying skills 203–6
  forming Interplanetary
    Society 48–9
  French penfriend 30–1
  General Magnetic
    company 175, 176, 201–3
  giving talks in German 93
  Golden Years 135, 141
  grandparents 1
  Great Yarmouth
    holiday 32, *33*
  headhunted by US
    universities 145–7
  Honeywell computer 86, *113*
  honorary degree from Leipzig
    University 223
  honorary doctorate from
    Cambridge University 210
  illness after first whole-body
    scan 126–7
  imaging group *c*.1980 *136*
  interviews and general
    discussion at Nobel
    awards 188, *188*
  Johnson and Johnson deal 143
  Johnson and Johnson
    invitation to US 145
  on Kent's Cavern 24–5
  knighthood 168–9, *170*, *171*
  Krakow visit 97
  language skills 28, 31, 93
  learning to drive 65
  Leipzig visit 96–7
  lino cuts 29
  meeting Damadian 196–9
  meeting future wife 45
  meeting Hounsfield 116, *117*,
    117–18
  meeting Swedish Royal
    family 191–2, *192*
  at military display in Trafalgar
    Square 17
  millennium celebration 182
  MRC grant application
    delayed by Andrew 119–20
  MRC Millennium
    Medal 209–10
  multi-pulse NMR 78–81,
    84–5, 86–7
  National Service 37–41
  Nobel lectures 187–8
  Nobel Prize awards
    ceremony 189–90, *190*
  Nobel Prize celebration
    speech 184
  Nobel Prize reception and
    dinner 191
  NRDC honour 162
  oil industry research 173–4
  at Oxford Polytechnic 41
  parents 1
  passport photograph *53*
  pedal-powered helicopter
    construction 78, *79*
  portrait painting by
    Shankland 206–7, *208*
  postdoctoral work at Urbana,
    Illinois 61, 64–75
  postgraduate studies 57–61,
    58
  on prediction of future Nobel
    Prize winning idea 88
  pre-Nobel Prize party 189
  presentation Ampère meeting,
    Holland 60
  presentation Ampère meeting,
    Krakow 109
  Pride of Britain Award 209
  printing interests 20, 29,
    32, 45
  professorship *ad hominem*
    offer from Oxford
    University 139–40

Index 237

projects in the Dudman's back garden 36
promoting Worthington's RCR Gold Medal award 221
prostate cancer 182–3
publications 60, 70–1, 80, 93, 110, 113, 127–8, 152–3, 178, 201, 203
Rank Prize award 142
receiving call-up papers for US Forces 65
receiving news of Nobel Prize award 183–4
responding to Damadian's article 185
rocketry interests 34–5, 36–7, 39
royalty issues 158–162, 178–9
at RPD Westcott 34–5, 41–5
sabbatical at Max Planck Institute, Heidelberg 85–6, 89–107
on safety of MRI 172
at Salvation Army Sunday School 96
Salzburg trip 53–6
schooling 11, 20, 27–8
in Sea Scouts 20–1
Siemens consultancy 132–3
sixty-fifth birthday conference 225, *225*
Southend visit 2, *4*
summer vacation jobs 51–2
summer vacations in US 71–3
tarantula encounter 72
tea room discussion with Garroway and Grannell 87–8
Teignmouth school trip 25
third year undergraduate project 49–51

Times Lifetime Achievement award 208
toy making and selling 22, 29
undergraduate prize 74
undergraduate studies at Queen Mary College 45, 47–51, *49*, *50*, 57
on university early retirement scheme 133–4
Upsala University visit 191
V-1 attacks 18–19
volunteering for first whole-body scan 125–6
war years 4–25
at Watts and Son 51–2, 53
wedding and honeymoon 63, *63*
Wolfson Foundation grant application 128
woodwork interest 11, 22
Mansfield, Philip 214
Mansfield, Rose Lillian 1–2, *4*, *33*, *63*, 98, 135
Mansfield, Sarah Jane 81, 91, 94, 99, 146, 185, *171*, *190*, 192, *192*
Mansfield, Sidney Albert (Sidney) 2, 3, 5, *5*, *6*, *10*, 21, *21*, 51, *63*, 98, 215–16
Mansfield, Sidney George 1, *4*, *33*, *63*, 81
Mansfield, Steven 98
Mansfield-Murphy, Charlotte 190
Mansfield-Murphy, Samantha 190
Mansfield–Issa equation 174, 175
Marian, Mrs 12, 25
Martin, Derek 48
Matthews, Paul *49*
Maudsley, Andrew *113*, 114, *121*, 132, 143, 194

Index

Max Planck Institute,
   Heidelberg 85–6, 89–107
medical imaging 137, 139, 141, 142, 152
Medical Research
   Council 118–19
Mehring, Michael 81, 85, 93
millennium celebration 182
Moiré fringes 52
Moore, Bill 111, *121*, 130, 134, *138*
Moran, Dick 68
Morgan, Irfona 49
Morris, Ken 42, *42*
Morris, Neville 42, *42*
Morris, Peter 84, *121*, 125, 126, 127–8, 131, *136*, 141, *168*, 172, 194, 225, *225*
Morrison shelter 14
Morrow, Bill *42*
movie MRI images 137
MRC Millennium
   Medal 209–10
Muller, Burton 200
multi-pulse NMR 78–81, 84–5, 86–7

Nairn, Sandy 206
National Research Development
   Corporation (NRDC) 155–8, 162
National Service 37–41
Needham, David 81
New Scientist 51
NMR apparatus, building with
   transistors 49–51
NMR diffraction
   experiments 102–7
NMR Imaging (Mansfield and
   Hahn) 217, 218
NMR Imaging in Biomedicine
   (Mansfield and Morris) 128

Nobel Prize
   arrival in Stockholm 185–6
   awards ceremony 189–90
   Department of Physics
      celebration 184
   interviews and general
      discussion 188, *188*
   Nobel lectures 187–8
   pre-awards party 189
   receiving news of
      award 183–4
   reception and dinner 191
   speculation 181–2
Nottingham University
   arrival and early work at 74, 75, 77–88
   early retirement
      scheme 133–5
   first MRI meeting 120–1
   Nobel Prize celebration 184
   royalty issues 158–62

oil industry research 173–4
Okera, Peera 49
Orchard Downs 64, 65
Ordidge, Roger *136*, 137, *138*, 169–70
Osbourne, Kenneth (Ginger) 30
Osbourne, Mr (science
   teacher) 28
Oxford Polytechnic 41
Oxford University, professorship
   ad hominem offer 139–40

Packer, Ken 173
paediatric imaging 137, 139
Papworth, Edward 49
Paradise, Helena 213
Parker, Malcolm 207, 223
Partain, Leon *138*
patents 111, 112–13, 114, 150–1, 155–63, 178

Index

Peckham Central School 27
penfriend 30–1
Percy, Bill 115
Peter (cat) 11–12
Pettie, Dan 57, 58
Pfeiffer, Harry 96, 223–4
phase compensation system 92
Philips Company 178–9
Pines, Alexander 68, 81
Pipe, Mr 28
Pohost, Gerry 146
portrait 206–7, *208*
Powles, Jack G. 48, 49, 51, 59, 60, 61, 70, 74, 80, 203, *204*, 224–6
Powles, Jill 225
Pride of Britain Award 209
printing 20, 29, 32, 45
prostate cancer 182–3
publications 60, 70–1, 80, 93, 110, 113, 127–8, 152–3, 178, 201, 203
pulsed spectrometer 58–9
Pykett, Ian *121*, 125, 135, *136*, 141, 142, 143, 194

Queen Mary College 45, 47–51, 57

Radda, George *138*, 139
Randall, Ed *204*
Rank Prize 142
reflection symmetry 82, 84–5
Richards, Ken 81
Richards, Rex 119, 139
Ridge, Joe 36
Riedel, Pappa *42*
Ringertz, Professor 220
Roads, Joan *42*
Rocket, The (journal) 48–9
Rocket Propulsion Department, Westcott 34–5, 41–5

rocketry 34–5, 36–7, 39
Rockets (Ley) 133
Roemer, Peter 158
rotating frame relaxation times 67–8
Rowland, Cecil Henry Coysh (Uncle) 8–11, *10*, 25–6
Rowland, Florence Susan (Auntie) 8–11, *10*, 25, 26
Rowland, Ted 61
Royal College of Radiology, Gold Medal 221
Royal Society Fellowship 120, 221
Royal Society meeting, Damadian's behaviour 199
royalty issues 158–162, 178–9
r-space 87, 109, 227
Rzedzian, Richard 135, *136*, 141, 142, 143

safety of MRI 172
Salvation Army 95–6
Salzburg trip 53–6
Sands, Ruth *49*
sandstone bore cores 173–4
Sarfas, Cliff *49*
Saunders, Rich *138*
Sayers, Martin, *49*
Schittenhelm, Dr 132, 133
Schmidt, Rakete 133
Scholls, Jeremy 160, 161
schooling 11, 20, 27–8
Schrieffer, John Robert 68
Schweitzer, Dieter 91, 96
Schwetzingen apartment 90
Scott, Dr 145
Scott, Kate 134
Sears, Cyril *49*
Sea Scouts 20–1
sensitive point imaging 111
Sevenoaks, evacuation to 4–7

240 Index

Shankland, Stephen 206–7, *208*
Shoulders, Alan 42
Shurblom, Rolf 191
Sidjon's 216
Siemens consultancy 132–3
Silvia, Queen of
 Sweden 192, *192*
Skinner family (John, Miriam
 and Arthur) 9–10, *10*,
 22, 52
Slarke, Norman 148
slice selection 111–14, 193–4
Slichter, Anne 223
Slichter, Charles Pence 61, 64,
 67, 68, 146, 210,
 223–4, *224*
Slichter, Nini 64
Smith, Arthur 215
Smith, Captain 40
Smith, E. T. B. *42*
Smith, Frank *138*
Smith, J. R. *42*
Smith, Professor 169, *170*
Smoothy, Steve 205
solid echoes 59–60, 70, 81–3
Southend 2, *4*
Spalding, Eric *42*
'Spider' 44
Spieß, Hans 91, 96
spin echoes 59, 81–3, 217
spin locking 83
sputnik sighting 56
Squire, Sarah 210
Stalker, Dennis 81, 102
Stapleton, Len 205
Stehlik, Dieter 91, 96
Stehling, Michael 140–1, 151
Stone, Jack *49*
Strakers apprenticeship 32
Strange, Annette 226
Strange, John 60, 70, 74–5, 203,
 204, 226

Sulston, John 209
Summerfield, Arthur *49*
Sutherland, Ruth 155, 156, 158

tarantula encounter 72
Technicare 143, 156, 157
Teignmouth school trip 25
Thatcher, Margaret, 168, *168*
Tiller, Mr 27
Times Lifetime Achievement
 award 208
Tompa, Kalmàn 97
Torquay, evacuation to 7–17,
 20–5
Torre Abbey (Meadow) 16
toy making and selling 22, 29
transistors 50
Türm, Frau 91
Turner, Jack 1, *3*
Turner, Margaret 1, *2*
Turner, Robert 150, 178

Upsala University 191
Urbana, Illinois 61, 64–75,
 145–6

V-1 17, 18–19
V-2 20
Vennart, Bill *121*
Volume, Mr 27
Vorderman, Carol 209

WaHuHa sequence 84
Wallace, George 146
Wallis, Mavis 52–3
Wallis, Roy 39, 41, 47, 48, 52–3
Ward, Philip 181
Ware, Donald 78
Watcombe Potteries 12
Watts and Son 51–2, 53
Waugh, John 80, 81, 84, 85,
 109, 127

Weedon, Basil 134, 159, 160, 161, 162, 167
Westcott, RPD 34–5, 41–5
Whizzer, The (comic) 29
whole-body imaging 122–6, 137
Williams, John 148
Willis, Peter 209
Winston-Salem conference 137, *138*
Wisconsin University 145–6

Wolfson Foundation grant 128, 129
Wood, Sir Martin 139
woodworking 11, 22
Worsfield, David *49*
Worthington, Brian *138*, 181, 184, 218–21

Young, Ian 52, 129, *138*, 199

Zimmerman, Herr 91